ISBN 978-1-5285-2321-9
PIBN 10901050

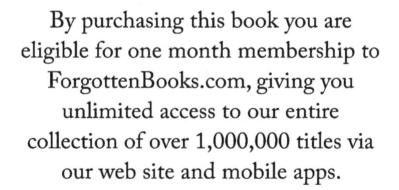

English
Français
Deutsche
Italiano
Español
Português

www.forgottenbooks.com

Mythology Photography **Fiction**
Fishing Christianity **Art** Cooking
Essays Buddhism Freemasonry
Medicine **Biology** Music **Ancient**
Egypt Evolution Carpentry Physics
Dance Geology **Mathematics** Fitness
Shakespeare **Folklore** Yoga Marketing
Confidence Immortality Biographies
Poetry **Psychology** Witchcraft
Electronics Chemistry History **Law**
Accounting **Philosophy** Anthropology
Alchemy Drama Quantum Mechanics
Atheism Sexual Health **Ancient History**
Entrepreneurship Languages Sport
Paleontology Needlework Islam
Metaphysics Investment Archaeology
Parenting Statistics Criminology
Motivational

MARYLAND GEOLOGICAL SURVEY

KENT COUNTY

MARYLAND

GEOLOGICAL SURVEY

KENT COUNTY

BALTIMORE
THE JOHNS HOPKINS PRESS
1926

ADVISORY COUNCIL

RAYMOND A. PEARSON - - - - - Executive Officer
PRESIDENT UNIVERSITY OF MARYLAND

FRANK J. GOODNOW - - - - - - Ex-officio Member
PRESIDENT JOHNS HOPKINS UNIVERSITY

ROBERT W. WILLIAMS - - - - - - - Baltimore

JOHN B. FERGUSON - - - - - - - Hagerstown

SCIENTIFIC STAFF

EDWARD BENNETT MATHEWS - - - - - STATE GEOLOGIST
SUPERINTENDENT OF THE SURVEY

EDWARD W. BERRY - - - - - ASSISTANT STATE GEOLOGIST

B. L. MILLER - - - - - - - - - - - GEOLOGIST

J. T. SINGEWALD, JR. - - - - - - - - - GEOLOGIST

Also with the coöperation of several members of the scientific bureaus of the National Government and Carnegie Institution.

LETTER OF TRANSMITTAL

To His Excellency ALBERT C. RITCHIE, *Governor of Maryland,*

Sir:—I have the honor to present herewith a report on The Physical Features of Kent County. This volume is the eighth of a series of reports on the county resources, and is accompanied by large scale topographical, geological, and agricultural soil maps. The information contained in this volume will prove of both economic and educational value to the residents of Kent County as well as to those who may desire information regarding this section of the State. I am,

Very respectfully,

EDWARD BENNETT MATHEWS,

State Geologist.

JOHNS HOPKINS UNIVERSITY,
BALTIMORE, *February, 1926.*

CONTENTS

ILLUSTRATIONS

PREFACE

This volume is the eighth of a series of reports dealing with the physical features of the several counties of Maryland.

The *Introduction* contains a brief statement regarding the location and boundaries of Kent County together with its chief physical characteristics.

The *Physiography of Kent County*, by Benjamin L. Miller, comprises a discussion of the surface characteristics of the county, together with a description both of the topographic forms and of the agencies which have produced them.

The *Geology of Kent County*, by Benjamin L. Miller, deals with the stratigraphy and structure of the county. An historical sketch is given of the work done by others in this field to which is appended a complete bibliography. Many stratigraphical details are presented, accompanied by local sections.

The *Mineral Resources* of Kent County, by Benjamin L. Miller, deals with the economic possibilities of the various geological deposits of the county. Those which have been hitherto employed are fully discussed, and suggestions are made regarding the employment of others not yet utilized.

The *Soils of Kent County*, by Jay A. Bonsteel, contains a discussion of the leading soil types of the county and their relation to the several geological formations. This investigation was conducted under the direct supervision of Professor Milton Whitney, Director of the Bureau of Soils of the U. S. Department of Agriculture.

The *Climate of Kent County*, by Roscoe Nunn, is an important contribution to the study of the climatic features of the county. Mr. Nunn had the benefit of a manuscript prepared some years ago by Mr. Wm. H. Alexander when Section Director in Baltimore of the U. S. Weather Bureau, and also Meteorologist of the Maryland State Weather Service.

The present report has been entirely rewritten and is based upon the more extended meteorological records now available.

The Hydrography of Kent County, by N. C. Grover, gives a brief account of the water supply of the county, which, as in the case of the other Coastal Plain counties, afford but little power for commercial purposes. The author of this chapter, formerly the Director of the U. S. Reclamation Service, is now chief of the Division of Hydrography of the U. S. Geological Survey.

The Magnetic Declination in Kent County, by L. A. Bauer, contains much important information for the local surveyors of the county. Dr. Bauer has been in charge of the magnetic investigations since the organization of the Survey and has already published two important general reports upon this subject. He is the Director of the Department of International Research in Terrestrial Magnetism of the Carnegie Institution.

The Forests of Kent County, by F. W. Besley, is an important contribution and should prove of value in the further development of the forestry interests of the county. Mr. Besley is State Forester of Maryland.

The State Geological Survey desires to extend its thanks to the several national organizations which have liberally aided it in the preparation of several of the papers contained in this volume. The Director of the U. S. Geological Survey, The Chief of the U. S. Weather Bureau, and the Chief of the Bureau of Soils of the U. S. Department of Agriculture have granted many facilities for the conduct of the several investigations and the value of the report has been much enhanced thereby.

THE

PHYSICAL FEATURES

OF

KENT COUNTY

THE PHYSICAL FEATURES OF KENT COUNTY

INTRODUCTION

Kent County lies between the parallels of 39° 1' and 39° 23' north latitude and between the meridians of 75° 46' and 76° 17' west longitude. It forms a part of the Eastern Shore of Maryland and has an area of 281 square miles. The county is bounded by water on the north, west, and south sides; it is separated on the north from Cecil County by the Sassafras River; the waters of Chesapeake Bay wash its shores on the west side; and on the south Chester River separates it from Queen Anne's County. As the estuaries of the Sassafras and Chester rivers extend almost to the Delaware line the county is bounded on three sides by navigable water. On the east side the Delaware-Maryland line, which was surveyed by Mason and Dixon about 1765, separate it from the Delaware counties of Kent and Newcastle. The extreme length of the county measured from the Delaware line to the extreme southwest corner is about 40 miles while the average width is less than 10 miles. This means that scarcely any part of the county is more than five miles distant from navigable water. In the early history of the region these streams played a very important part in the development of the region and by means of them the inhabitants of the county were brought in close communication with the residents of the western shore counties of St. Mary's and Anne Arundel.

The history [1] of Kent County is a long and interesting one. The county was named for Kent County, England, the name, however,

---

[1] For a more extended account see "The Counties of Maryland, their Origin, Boundaries, and Election Districts" by Edward B. Mathews, Md. Geol. Survey, vol. vi, pp. 419-572, 1906.

being first applied to Kent Island where a trading post was established by William Claiborne in 1631. The earliest known reference to Kent County was made in 1642 and it is supposed that at that time the county was intended to include all the settlements on the eastern shore of Chesapeake Bay just as St. Mary's County included the entire inhabited portion of the western shore. In 1659 part of the county on the north was taken off to form Baltimore County and in 1662 Talbot County was organized and a large area on the south was set off from Kent. Baltimore County probably included a portion of what is now Kent County, viz., the settlements on the south shore of the Sassafras River. Later in 1674 when Cecil County was erected out of Baltimore County, this same region was made a part of the new county. In 1671 Kent Island was removed from Kent County and annexed to Talbot though a considerable part of what now constitutes Queen Anne's County still formed a part of Kent. In 1707 the boundaries of the county were established approximately as they are at the present time though the eastern boundary was not definitely fixed until 1750 and the survey was not made until about 15 years later.

The indefinite character of the boundaries of the county in early colonial times is explained by the history of its development. The whole region was covered with dense forests and with no roads crossing the peninsula, all the first settlements were made along the larger water courses and communication was effected solely by water. Consequently the inhabited and unexplored divides made more satisfactory boundary lines between the counties than did the streams which divided settlements whose interests were more closely united. It was not until almost a century after the first settlement of the region that the divide between the Chester and Sassafras rivers became inhabited sufficiently to unite the settlers on the two sides of the county.

Agriculture is the principal occupation of the inhabitants of Kent County and has been during almost the entire period since its earliest settlement by the white men. Claiborne's first settlement on Kent Island, prior to the founding of St. Mary's City, was mainly for the purpose of trading with the Indians but with the increase of settlers, farming soon became the chief pursuit. Prior to the coming of the white men the region was inhabited by Indians as is evidenced by the great accumulations of oyster shells found in various places along Chesapeake Bay. These shell heaps mark the sites of old Indian villages and usually are found on rather elevated points commanding good views of the surrounding country. One on the high land near Howell Point is probably the most northerly Indian kitchen midden of Chesapeake Bay.

During the time that has elapsed since the settlement of the region probably every acre of land in the county has been under cultivation and at the present time there is very little land that is not being cultivated. No doubt there are many farms throughout the county that have been under practically continuous cultivation for over 250 years.

There are no large towns in the county, Chestertown, the county seat, being the largest; Millington, Sassafras, Galena, Kennedysville, Stillpond, and Fairlee are small hamlets situated in the midst of thriving farming communities and supported by them; while Betterton, Tolchester Beach, and Rockhall are popular summer resorts on Chesapeake Bay much frequented by residents of Baltimore.

Two branches of the Philadelphia, Baltimore, and Washington Railroad enter the county, bringing the region into direct communication with the principal cities of the Atlantic seaboard, while several lines of steamers ply between various points along the Chester and Sassafras rivers and Baltimore and other points on Chesapeake Bay, while the boats between Baltimore and Philadelphia make regular stops at Betterton. In this way the whole county is in close communication with adjoining regions and facilities for transportation of the farm products to market are excellent. Much grain is shipped to Baltimore at a minimum of expense on small sailing vessels that are able to pass far up the Chester and Sassafras river estuaries.

In recent years the State Roads Commission has built improved highways connecting Chestertown, the county seat, with Rock Hall and Tolchester on the Chesapeake Bay with the spur line into Crosby on Lankford Bay. To the north an improved roadway runs to Georgetown by way of Galena, connecting with the Cecil County system. A spur line runs from this road into Betterton on the Sassafras River. An improved highway crosses the county from Galena to Millington where it connects with the Queen Anne's road system. Another good road extends from Galena to Lambson. To the south a highway is built paralleling the water front of the Chester River from Chestertown to Pomona. Chestertown is connected with the county seat of Queen Anne's County by an improved highway.

DEVELOPMENT OF KNOWLEDGE CONCERNING THE PHYSICAL FEATURES OF KENT COUNTY, WITH BIBLIOGRAPHY

BY

BENJAMIN L. MILLER

INTRODUCTORY.

Since 1608 when Captain John Smith explored the upper portion of the Chesapeake Bay the Coastal Plain of Maryland has attracted the attention of explorers, travelers, and geologists, many of whom have published their observations.

In this review no attempt is made to include all who have written on the geology of the region but only those who have rendered most service in advancing our knowledge of the geology of the area, consequently investigators are mentioned rather than collaborators. The bibliography which follows gives the names of both.

Maclure in 1809 was the first geologist in this country to attempt to separate the different kinds of rocks on the basis of lithologic differences. These divisions were termed formations. He noted the wide difference in the characters of the rocks composing the Piedmont Plateau and the Coastal Plain and on the basis of these differences established two formations. He called the crystalline rocks of the Piedmont Plateau the "Primitive formation," and the unconsolidated deposits of the Coastal Plain the "Alluvial formation." His conclusions, accompanied by a colored geologic map on which these divisions were represented were published several times, but most fully in 1817. The work of Maclure served as a great incentive to geological research in this country outlining as it did the methods of work which have been followed since his time and which have yielded such important results.

Ducatel, State Geologist of Maryland from 1834 to 1840, was the first person to publish any definite information of value concerning the geology of Kent County. In his first report, published in 1834, he refers to the fossiliferous deposits at "Frederick ferry" on the Sassafras River and three miles below Chestertown on the Chester River and discusses the economic value of the marls of the Coastal Plain. He elaborates upon the same subjects in his report for the year 1835 and in his 1836 report (published in 1837), he again calls attention to the shell and greensand marls occurring within the region, which he thinks might prove valuable as fertilizers. In his 1837 report (published in 1838), he discusses the physiography and geology of Kent County in a more detailed manner.

In 1892 Clark in his article on "The Surface Configuration of Maryland," gave many facts pertaining to the topography of Kent County. In the following year Williams and Clark brought together in the volume, "Maryland, its Resources, Industries, and Institutions," all that was then known in regard to the physical features, geology, and mineral resources of the State, while in 1897 Clark in Volume I of the Maryland Geological Survey, and in 1906 Clark and Mathews, contributed more detailed reports on the same subjects. These reports contain brief descriptions of all the geological formations of the State and county then recognized, and much information regarding the physical features and economic resources.

Another important publication is "The Dover Folio" of the United States Geological Survey by Miller. The area described includes the greater portion of Kent County and is the most complete work on the general geology of the region published up to the present. The Systematic Reports, especially those on the Upper Cretaceous, Eocene, Miocene, Pliocene and Pleistocene describe the formations and their fossils in greater detail.

The work on the various geological formations found in Kent County may be summarized as follows:

THE UPPER CRETACEOUS.

In 1830 Morton described some fossils from the greensand strata of the Chesapeake and Delaware Canal and stated that they were pre-Tertiary in age. Eaton had previously claimed that the New Jersey greensands belonged to the Tertiary. Ducatel in 1836 was the first writer to mention definite localities in Kent County where greensands occur. He included all the Potomac, Marine Cretaceous, and Eocene deposits of the county in his "ferruginous sand formation." The lithologic characteristics of the strata are accurately described and mention is made of the occurrence of several Cretaceous fossils in Kent County. The strata are correlated with the greensands of New Jersey which are said to belong to the Secondary period.

In 1865 Conrad described the lignitic beds now included in the Magothy formation, and supposed that they represented the base of the Eocene. Darton in 1893 proposed the differentiation of the Magothy formation and described its lithologic characteristics and distribution. Roberts in 1895 gave detailed descriptions of several localities in Kent County where he obtained Cretaceous fossils.

In 1895 Clark presented a paper before the Geological Society of America in which the Upper Cretaceous deposits of the entire state were discussed. A map showing the distribution of the strata accompanied the article. White, in 1891 in a correlation bulletin of the United States Geological Survey, and Clark in 1897 in Volume I of the Maryland Geological Survey summed up all existing knowledge concerning the Upper Cretaceous formations of the State in which there are references to Kent County localities. Somewhat more detailed information covering a large portion of the county is contained in the Dover Folio by Miller published in 1906.

The Eocene.

The Eocene deposits of Kent County have received little attention in the published literature. Only two localities have been mentioned many times. These are the fossiliferous deposits at Fredericktown on the Sassafras River, and a few miles below Chestertown on the Chester River. The fossiliferous greensand deposit at Fredericktown has generally been correlated with the Cretaceous greensand deposits of New Jersey because of the presence of fossils formerly supposed to be confined to the Cretaceous. Ducatel in his 1834 report referred to these deposits and in his succeeding report correlated them with the New Jersey greensand. The same writer mentioned the fossiliferous strata along the Chester River, below Chestertown, and in his 1837 report correctly referred those deposits to the Tertiary period.

In 1834 Lea first applied the term Eocene to the Lower Tertiary deposits of America but it was not until Tyson published his annual report in 1860 that it was definitely stated that Eocene strata occur on the Chester River. From that time until 1901 the various published articles on the Eocene were mainly discussions of the correlation of the Northern Atlantic Coastal Plain strata with those of the Gulf states and Europe. Conrad, Heilpin, Uhler, Darton, and Clark contributed to these discussions.

Since 1888 Clark has been the principal investigator of the Eocene deposits of Maryland and in several of his published articles he has referred to the Eocene strata of Kent County. He proposed the classification of the Eocene adopted in this report. The most complete article is by Clark and Martin in the Eocene volume of the Maryland Geological Survey published in 1901. In this volume the fossils of the region are fully described by specialists and each recognized species is illustrated. In 1906 Miller in the Dover Folio of the United States Geological Survey also gave much detailed information concerning the Eocene deposits of Kent County.

THE MIOCENE.

The Miocene strata, although extremely fossiliferous elsewhere in Maryland, are practically barren of fossils in Kent County, consequently there are few references in the literature to the Miocene deposits of the region. The Miocene deposits of Kent County present few good exposures and have a rather limited distribution in the southeastern part of the county. In 1842 Conrad, who called the Miocene the Medial Tertiary, referred to deposits of this age in the vicinity of Chestertown and this is almost the only reference to the Miocene deposits of the county until within recent years. W. B. Rogers in 1836 was the first to announce the presence of Miocene deposits in Maryland. Conrad later accepted Roger's conclusion and between 1830 and 1869 laid the basis for exact correlation through his paleontological work on the molluscan fauna while Bailey, Ehrenberg and Johnston studied the microscopic forms which are so abundant in the diatomaceous earth of the Calvert formation.

THE PLEISTOCENE.

Although the Pleistocene deposits cover such a large portion of the Coastal Plain they received little attention by the early geologists. This is mainly due to the fact that except in a very few places, fossils are either extremely rare or entirely absent. Occasional mention is made of the surficial sands and gravels but the references are brief and indefinite. Chester in 1885 published the results of his study of the sands and gravels of the peninsula of Delaware and the Eastern Shore of Maryland. He attributed their origin to the Delaware River while the large boulders were said to have been carried by icebergs. McGee in 1887 and 1888 published several papers on the Columbia deposits in which he described the deposits in detail and gave many sections along the shores of Kent County. He described the deposits as constituting a series of deltas and terraced littoral deposits. The ice-borne boulders brought down by the Susquehanna are said to have been fifty times as large as those carried at the present time.

Darton in articles published in 1891, 1893, and 1901 made valuable contributions to our knowledge of these formations. In an article published in 1901 Shattuck described the gravel deposits of the North Atlantic Coastal Plain, reviewed former ideas and classifications of these late formations, and proposed the classification adopted in this report. The latest and most complete discussion is contained in a recent volume, issued by the Maryland Geological Survey in 1906, on the Pliocene and Pleistocene Formations of Maryland. It contains a full discussion of the deposits and also the fauna and flora which they contain

BIBLIOGRAPHY

1624.

SMITH, JOHN. A Generall Historie of Virginia, New England, and the Summer Isles, etc. London, 1624. (Several editions.)

This work contains many interesting notes on the physiography of Chesapeake Bay and its tributaries, and briefly described the clays and gravels along their shores. For a reproduction and discussion of Smith's map see Md. Geol. Surv., Vol. II, pp. 347-360.

1817.

MACLURE, WILLIAM. Observations on the Geology of the United States of America, with some remarks on the effect produced on the nature and fertility of soils by the decomposition of the different classes of rocks. 12 mo. 127 pp. 2 pls. Philadelphia, 1817. Is an elaboration of an article published in 1809 in Trans. Amer. Phil. Soc. O.S., Vol. VI, pp. 411-428. Republished in Trans. Amer. Phil. Soc. N.S., Vol. I, 1818, pp. 1-91.

This work is classic as it was the first attempt to treat the geology of the entire country and it contains the first published geological map of the United States. In this work the whole of the Coastal Plain sediments constitute the "Alluvial" formation and the Piedmont Plateau the "Primitive."

1824.

FINCH, JOHN. Geological Essay on the Tertiary Formations in America. (Read before Acad. Nat. Sci., Phila., July 15, 1823.)

Amer. Jour. Sci. Vol. VII. pp. 31-43.

Objection is made to the term "Alluvial formation" of Maclure and others on the ground that the deposits are for the most part not of alluvial origin and also that, as used, it includes a number of distinct formations that can be correlated with the "newer secondary and tertiary formations of France, England, Spain, Germany, Italy, Hungary, Poland, Iceland, Egypt, and Hindoostan." The writer makes some provisional correlations with European formations which are now known to be incorrect. He admits, however, that the data are insufficient for accurate correlation.

1826.

PIERCE, JAMES. Practical remarks on the shell marl region of the eastern parts of Virginia and Maryland, etc., extracted from a letter to the Editor.

Amer. Jour. Sci., Vol. XI, pp. 54-59, 1826.

Mentions the occurrence of shell marl of marine origin in the "alluvial" district of Maryland on both sides of Chesapeake Bay and discusses its value as a fertilizer in the renovation of exhausted soils.

1830.

MORTON, SAMUEL G. Synopsis of the Organic Remains of the Ferruginous Sand Formation of the United States, with geological remarks.

Amer. Jour. Sci., Vol. XVII, pp. 274-295; Vol. XVIII, pp. 243-250, 1830.

The writer describes fossils from the greensand marls of New Jersey, from the Deep Cut of the Chesapeake and Delaware Canal, and from Maryland. The author contends that the greensands are pre-Tertiary in age and should be correlated with the Lower Chalk of England. Eaton had claimed that the beds were of Tertiary age.

1834.

DUCATEL, J. T. and ALEXANDER, J. H. Report on the Projected Survey of the State of Maryland, pursuant to a resolution of the General Assembly. 8 vo. 39 pp. Annapolis, 1834. Map. Several editions.

Amer. Jour. Sci., Vol. XXVII, 1835, pp. 1-39.

Fossiliferous deposits occurring at "Frederick ferry" on the Sassafras River and three miles below Chestertown on the Chester River-are described and the statement is made that "these spots may perhaps be indicated as the commencement of the fossiliferous deposits of the Eastern shore of Maryland."

1835.

CONRAD, J. A. Observations on the Tertiary Strata of the United States.

Amer. Jour. Sci., Vol. XXVIII, pp. 104-111, 280-282.

He considers the Miocene absent in this region, the Older Pliocene resting directly upon the Eocene. The beds containing *Perna maxillata* are referred to the Older Pliocene and the St. Mary's river beds to the Medial Pliocene.

DUCATEL, J. T. and ALEXANDER, J. H. Report on the New Map of Maryland, 1834. Annapolis, 1835(?). 8 vo. 59+i pp. Two maps and one folded table. Contains Engineer's and Geologist's Reports which were also issued separately. Md. House of Delegates, Dec. Sess. 1834.

Oyster shell heaps near Worton Point are mentioned and their value as fertilizers suggested. Shell marl is described and Ducatel says that he believes it underlies most of the Eastern shore though not exposed south of the Choptank river. He says that it has a dip of 5° to the southwest, while the surface of the marl undulates.

1836.

DUCATEL, J. T. and ALEXANDER, J. H. Report on the new Map of Maryland, 1835. 8 vo. 84 pp. Maps. Annapolis, 1836.

Md. Pub. Doc., Dec. Sess. 1835.

Engineer's Report, pp. 1-34, Geologist's Report, pp. 35-84.

Both reports also published separately.

Ducatel states that greensand of the age of "the New Jersey marl has been satisfactorily ascertained to occur at the head of the Sassafras River in Kent and Cecil counties and seems to underlie nearly the whole of Kent County. It forms a part of the ferruginous sand formation." The greensand at one place on the Sassafras River is said to be filled with shells of *terebratulae*. The "ferruginous sand formation" is said to be "very variable, consisting of local and circumscribed deposits of clay, sand, and gravel, most of them highly ferruginous and varying in color from deep red, yellow, gray and green, to black and bluish black." Besides the greensand the "micaceous black sand" is described in detail. It is said to contain in many places, iron pyrite, spicules of selenite, and fossils, usually in the form of casts. The best

FIG. 1.—VIEW OF THE CHESTER RIVER AT MILLINGTON.

FIG. 2.—VIEW SHOWING BLUFF CUT IN CRETACEOUS AND PLEISTOCENE DEPOSITS BY
CHESAPEAKE BAY AT BETTERTON.

preserved fossil is *Ostrea falcata*. At the head of Churu Creek the material was found to contain "a species of *Turritella*, the *Cucullea vulgaris* of Dr. Morton, claws of Crustacea, teeth of a saurian animal, fish bones, wood perforated by marine insects, etc." The micaceous black sand is also noted at Fairlee where it contains iron pyrites and selenite, and is overlain by a gravel and boulder deposit. Its value as a fertilizer is doubtful, as both beneficial and injurious results have been given by farmers who have applied it to their lands. Analyses of greensand, micaceous black sand, clay, ochre, siliceous sand, and shell marl are given. (p. 83).

1837.

DUCATEL, J. T. Outline of the Physical Geography of Maryland, embracing its prominent Geological features.

Trans. Md. Acad. Sci. and Lit., Vol. I, Pt. I, pp. 24-55. 1837. With map.

A general description of the physiography and geology of the entire state is given with many details of local features. It is a general summary of information previously published in various places. Mention is made of the covering of boulders and coarse gravel near the inner edge of the Secondary (Cretaceous) rocks while farther out the sands and clays of the Secondary and Tertiary formations are said to be uncovered.

The secondary rocks are said to cover practically all the county except along the Chester River. In the greensand the fossils are *Terebratulae* and *Gryphaea vomer* and "in the micaceous black sand there have been found the *Exogyra*, *Ostrea falcato*, casts of *Cucullaea mortonii*, fragments of Ammonites, the tooth of a saurian reptile, claws of a species of crab, lignites, with other undetermined organic bodies, and in some localities pyrites and crystals of selenite."

1838.

DUCATEL, J. T. Annual Report of the Geologist of Maryland. 1837. Annapolis, 1838. 8 vo. 39 pp., 2 maps.

Md. Pub. Doc., Dec. Sess. 1837.

A good general description of the physiographic features of the county is given. The soils of the different portions are described and the adaptability to various crops discussed. "In reference to its geological constitution, the northern and middle portions of the county are based upon deposits of the secondary period, referable to what in our country has been termed the ferruginous-sand formation, and embracing extensive beds of greensand containing as characteristic fossils *terebretula* and *gryphaca*, and beds of a micaceous black sand with *belemnites, ammonites, exogyrae*, etc. The superincumbent deposits of clay, sand and gravel, that occasionally present themselves, have very little depth, and belong doubtless to a much more recent epoch, which it is difficult to assign with precision. The only fossil known to have been found in them, is the grinder of a mastodon. They are probably of diluvial origin."

Deposits belonging to the Tertiary period are said to occur in the southwest portion of the county along the Chester River. At Farley there is a lignitiferous clay of this age containing nodules of pyrite and "detached and grouped crystals of selenite." It is overlain by "a thick stratum of boulder and gravel composed of coarse and fine-grained sandstone, green stone, micaceous and argillaceous slates, quartz-rock and quartz, from several hundred pounds weight down to ordinary sized gravel, the whole covered by a clayey-loamy soil upwards of three feet in depth." The writer advises the use of greensand, black micaceous sand and oyster shells from the Indian oyster shell heaps as fertilizers and cites instances where they have been used with beneficial effects.

Bog-iron ore of good quality is reported at the head of a branch of Worton Creek on the farm of Mr. Levy Wroth.

1842.

CONRAD, T. A.   Observations on a portion of the Atlantic Tertiary Region, with a description of new species of organic remains.

2d Bull. Proc. Nat. Inst. Prom. Sci., 1842.   pp. 171-192.

The Miocene and Eocene are said to not be connected by a single fossil common to both periods while three forms found in the Upper Secondary are found in the Eocene.

.The Medial Tertiary (Miocene) is said to appear near Chestertown.

1843.

DUGATEL, JULIUS T.   Physical History of Maryland.

Abstract, Proc. Amer. Phil. Soc., Vol. III, 1843, pp. 157-158.

"The Eastern Shore is shown to consist of something more than arid sand-hills and pestilential marshes; and the Western Shore not to depend exclusively upon the rich valleys of Frederick and Hagerstown for its supplies."

1850.

HIGGINS, JAMES.   Report of James Higgins, M. D., State Agricultural Chemist, to the House of Delegates.   8 vo.   92 pp.   Annapolis, 1850.

. Contains detailed descriptions and many analyses of the various kinds of soils found on the Eastern Shore of Maryland.   The greensand and shell marl deposits of the counties lying north of the Choptank River are discussed at length and many references made to localities in this county where they occur.

1852.

FISHER, R. S.   Gazetteer of the State of Maryland compiled from the returns of the Seventh Census of the United States.   New York and Baltimore, 1852.   8 vo., 122 pp.

Contains numerous brief descriptions of the geography and geology of different portions of the State.

1860.

TYSON, PHILIP T. First Report of Philip T. Tyson, State Agricultural Chemist, to the House of Delegates of Maryland, Jan. 1860. 8 vo. 145 pp. Maps. Appendix. Mineral Resources of Md. 20 pp. Annapolis, 1860.

The report is accompanied by a colored geological map which shows the distribution of the various formations. The Coastal Plain formations represented are the Cretaceous, Tertiary, and Post-Tertiary, while the iron-ore clays of the Cretaceous are separated from the other Cretaceous deposits. A brief description of each formation is given.

Greensand marl of Eocene age is reported to occur along the Chester River.

1867.

HIGGINS, JAMES. A Succinct Exposition of the Industrial Resources and Agricultural Advantages of the State of Maryland. 8 vo., 109+III pp.

Md. House of Delegates, Jan. Sess., 1867, (DD).

Md. Sen. Doc., Jan. Sess., 1867, (U).

Contains a description of the soils and physiographic features of each of the counties of the State.

1883.

SMOCK, J. C. The Useful Minerals of the United States.

Min. Resources of the U. S., 1882. Washington, 1883. pp. 690-693.

The following minerals are reported from this county: greensand marl from head of Sassafras River, and lignite occurring sparingly in clay.

WILBUR, F. A. Marls.

Mineral Resources U. S., 1882. Washington, 1883, p. 522.

Greensand marls of Cretaceous age said to occur in Kent, Cecil, and Prince George's counties.

1884.

CHESTER, FREDERICK D. The quaternary Gravels of Northern Delaware and Eastern Maryland, with map.

Amer. Jour. Sci., 3d ser., Vol. XXVII, 1884, pp. 189-199.

The author believes that the peninsula of Eastern Maryland and Delaware was covered with gravels, clay and sand brought down by the Delaware River during the Ice Age and deposited in an estuary.

1885.

CHESTER, FREDERICK D.   The gravels of the Southern Delaware Peninsula.

Amer. Jour. Sci., 3d ser., Vol. XXIX, 1885, pp. 36-44.

The gravels, sands, and clays of the entire peninsula of Eastern Maryland and Delaware are said to have been brought down by the Delaware River and spread out by estuarine and marine currents. In the northern part the materials were deposited in an estuary but in the southern part in the open ocean. Boulders carried by icebergs are found throughout the entire area, some of which are of large size.

1888.

MCGEE, W. J.   The Geology of the Head of Chesapeake Bay.

7th An. Report U. S. Geol. Surv., Washington, 1888, pp. 537-646.

(Abst.) Amer. Geol., Vol. I, 1887, pp. 113-115.

Contains a general discussion of the Potomac and Columbia deposits. Many sections along the Sassafras River are described in detail.

MCGEE, W. J.   The Columbia Formation.

Proc. Amer. Assoc. Adv. Sci., Vol. XXXVI, 1888, pp. 221-222.

The Columbia formation overlying unconformably the Cretaceous and Tertiary deposits of the Atlantic Coastal Plain is said to consist of series of deltas and terraced littoral deposits. It is said to pass under the terminal moraine to the northward. The Columbia materials are supposed to have been laid down during a period of glaciation long preceding the glacial epoch during which time the terminal moraine was formed.

———————————— Three Formations of the Middle Atlantic Slope.

Amer. Jour. Sci., 3d ser., Vol. XXXV, 1888, pp. 120-143, 328-331, 367-388, 448 466, plate II.

The three formations discussed are the Potomac, (now divided into four formations), the Appomattox (Lafayette), and the Columbia, (now divided into three formations). These are described in far greater detail than had ever been done before and the conclusions reached vary but little from the views held at the present time.

UHLER, P. R. Observations on the Eocene Tertiary and its Cretaceous Associates in the State of Maryland.

Trans. Md. Acad. Sci., Vol. I, 1888, pp. 11-32.

Many details concerning the distribution, lithologic characteristics, and fossil content of the Eocene and Cretaceous deposits of this county are given.

1889.

UHLER, P. R. Additions to Observations on the Cretaceous and Eocene formations of Maryland.

Trans. Md. Acad. Sci., Vol. I, 1889. pp. 45-72.

This paper contains many descriptions of Cretaceous and Eocene strata in this county together with a general description of these formations as represented in the entire state. A list is given of all Eocene fossils recognized up to that time.

1891.

CLARK, WM. B. Correlation Papers—Eocene.

Bull. U. S. Geol. Surv. No. 83. Washington, 1891. 173 pp. 2 maps.

(Abst.) Johns Hopkins Univ. Cir. No. 103, Vol. XII, 1893, p. 50.

Contains a discussion of all the literature concerning the Eocene of the United States published up to that time. The distribution and characteristics of the Maryland Eocene deposits are briefly described.

1892.

CLARK, WM. B. The Surface Configuration of Maryland.

Monthly Rept. Md. State Weather Service, Vol. II, 1892. pp. 85-89.

General summary of the physical features of the State.

SCHARF, J. THOMAS. The Natural Resources and Advantages of Maryland, being a complete description of all of the counties of the state and the City of Baltimore. Annapolis, 1892.

This paper contains general information concerning this county.

1893.

CLARK, WM. B.   Physical Features (of Maryland).

Maryland, its Resources, Industries, and Institutions.   Baltimore, 1893, pp. 11-54.

Contains short descriptions of the topography, climate, water supply, and water power of the different portions of the State.

DARTON, N. H.   The Magothy Formation of Northeastern Maryland.

Amer. Jour. Sci., 3d ser., Vol. XLV, 1893.   pp. 407-419.   Map.

The Magothy formation is differentiated from other Cretaceous strata with which the deposits had previously been included. The distribution and characteristics of the formation are discussed and many local details described. A map showing the distribution of the formation is given.

WHITNEY, MILTON.   Description of the Principal Soil Formations of the State (Maryland).

Maryland, its Resources, Industries, and Institutions.   Baltimore, 1893, pp. 181-211.

Contains descriptions of the soils of the State, their distribution, origin, and adaptabilities.

WHITNEY, MILTON.   The Soils of Maryland.

Md. Agric. Expt. Sta., Bull. No. 21, College Park, 1893.   58 pp. Map.

The principal soils of the State are described and their adaptability to different kinds of crops discussed. A map is given showing their general distribution.

WILLIAMS, G. H. and CLARK, W. B.   Geology of Maryland.

Maryland, its Resources, Industries, and Institutions.   Baltimore, 1893, pp. 55-83.

The different geological formations recognized at that time are briefly described. Several important Eocene and Cretaceous localities in this county are mentioned.

1894.

ANON.   General Mining News—Maryland.

Eng. and Min. Jour., Vol. LVIII, 1894, p. 61.

Note concerning a deposit of amber in the Cretaceous beds on the Bay Shore above Still Pond in Kent County.

DARTON, N. H. Artesian Well Prospects in Eastern Virginia, Maryland, and Delaware.

Trans. Amer. Inst. Min. Eng., Vol. XXIV, 1894. pp. 372-397, pls. I and II.

Contains a general description of the Atlantic Coastal Plain formations with records of some of the important artesian wells of eastern Virginia, Maryland, and Delaware, with a discussion of artesian water conditions in those areas.

MARYLAND STATE WEATHER SERVICE. The Climatology and Physical Features of Maryland.

1st Bien. Rep. Maryland Weather Service for years 1892-1893. Baltimore, 1894.

A general discussion of the topography, geology, soils, and climate of the state.

1895.

ROBERTS, D. E. Note on the Cretaceous Formations of the Eastern Shore of Maryland.

Johns Hopkins Univ. Circ. Vol. XV, 1895. pp. 16-17.

The Redbank formation of the Cretaceous is said to occur at Fredericktown (north end of bridge) where it contains the following fossils: *Ostrea larva*, Lam.; *Exogyra costata*, Say.; *Dentalium falcatum*, Con.; and *Turritella encrinoides*, Mort.

The Rancocas formation is said to occur on Jackson's Farm, Herring Creek, where it contains the fossils, *Terebratula harlani;* Mort. and *Gryphea vesicularis*, Lam.

1896.

CLARK, W. B. The Eocene Deposits of the Middle Atlantic Slope in Delaware, Maryland, and Virginia.

Bull. 141, U. S. Geol. Surv., 167 pp. 40 pl.

An exhaustive study of the Eocene in which the stratigraphy and paleontology of the deposits are discussed in detail.

DARTON, N. H. Artesian Well Prospects in the Atlantic Coastal Plain Region.

Bull. 138, U. S. Geol. Surv., 232 pp., 19 pls.

Contains a brief description of the Coastal Plain formation of the state with a discussion of their water bearing qualities. Records are given of many deep wells in this State but none from Kent County.

1897.

CLARK, W. B. Outline of the Present Knowledge of the Physical Features of Maryland, Embracing an Account of the Physiography, Geology, and Mineral Resources.

Md. Geol. Survey, Vol. I, 1897, pp. 141-228, pls. 6-13.

Contains a description of all the geologic formations of the State recognized at that time.

CLARK, W. B., (with R. M. BAGG and G. B. SHATTUCK). Upper Cretaceous Formations of New Jersey, Delaware and Maryland.

Bull. Geol. Soc. of America, Vol. 8, 1897, pp. 315-358, pls. 40-50.

Contains a full description of each of the marine Cretaceous formations of the Northern Atlantic Coastal Plain.

1898.

BAGG, RUFUS MATHER. The Occurrence of Cretaceous Fossils in the Eocene of Maryland.

Amer. Geol., Vol. 22, 1898, pp. 370-375.

A Cretaceous shell layer is reported to "occur on a branch of the Sassafras River called Swan Creek on Mr. Jacob's farm."

1899.

ABBE, CLEVELAND, JR. General Report on the Physiography of Maryland.

Maryland Weather Service, Vol. 1, Baltimore, 1899, pp. 41-216, pls. 3-19, figs. 1-20.

Contains a full description of the physiographic features of the State.

WOOLMAN, LEWIS. Artesian Wells in New Jersey.

Geol. Surv. of New Jersey, Annual Report for the Year 1898, pp. 59-144. Trenton, 1899.

Contains descriptions and records of four artesian wells at and near Rock Hall ranging in depth from 175 to 400 feet. A list of 40 species of diatoms determined by Charles S. Boyer from the well samples is also given.

1900.

ABBE, CLEVELAND, JR. The Physiographic Features of Maryland.

Bull. Amer. Bur. Geog., Vol. I, pp. 151-157, 242-248, 342-355, 2 figs. 1900.

A concise statement of the important physical features of each of the three physiographic provinces of the State.

WOOLMAN, LEWIS. Artesian Wells.

Geol. Surv. of New Jersey. Annual Report for the year 1899. pp. 53-139. Trenton 1900.

Contains a short description of an artesian well at Kennedyville. (p. 81).

1901.

BONSTEEL, JAY A. Soil Survey of Kent County, Md.

Field Operations of the Division of Soils, 1900. pp. 173-186, 1 map.

Contains descriptions of the various kinds of soils recognized in the county.

CLARK, W. B., with collaborators. Systematic Paleontology,— Eocene.

Md. Geol. Surv. Eocene. Balto., 1901, pp. 95-215, pls. 10-64.

Contains descriptions and figures of all Eocene fossils known to occur within the State.

CLARK, W. B. and MARTIN, G. C. Eocene Deposits of Maryland.

Md. Geol. Surv., Eocene. Balto., 1901, pp. 21-92, 14 pls.

Describes the general stratigraphic relations, distribution, characteristic origin of the materials, and the stratigraphic and paleontologic characteristics of the Eocene strata of the entire State.

SHATTUCK, GEORGE BURBANK. The Pleistocene Problem of the North Atlantic Coastal Plain.

Johns Hopkins Univ. Circ. Vol. XX, 1901, pp. 69-75.

Amer. Geologist, Vol. xxvii, 1901, pp. 87-107.

The views of McGee, Darton, and Salisbury concerning the Pleistocene deposits are summarized and compared with the writer's views. The wave-built terrace deposits are referred to four different formations, the Talbot, Wicomico, Sunderland, and Lafayette, the first three of which constitute the Columbia group. These formations are said to be separated by erosional unconformities.

1903.

RIES, HEINRICH.   The Clays of the United States East of Mississippi River.

U. S. Geol. Surv. Prof. Paper No. 11, pp. 134-149.   1903.

Describes the clay bearing formations of the county and gives analyses and physical characteristics of the most important clays.

1904.

CASE, E. C., EASTMAN, C. R., MARTIN, G. C., ULRICH, E. O., BASSLER, R. S., GLENN, L. C., CLARK, W. B., VAUGHAN, T. W., BAGG, R. M., JR., HOLLICK, ARTHUR, and BOYER, C. S.   Systematic Paleontology of the Miocene Deposits of Maryland.

Md. Geol. Surv., Miocene, pp. 1-508, pls. 10-135.   Balto., 1904.

Contains descriptions and illustrations of all Miocene fossils recognized in Maryland up to that time.

CLARK, WILLIAM BULLOCK.   The Matawan Formation of Maryland, Delaware, and New Jersey, and its relation to overlying and underlying formations.

Amer. Jour. Sci., 4th ser., Vol. 18, pp. 435-440, 1904.

Johns Hopkins Univ. Circ., 1904, No. 7, pp. 28-35.

The Matawan formation as it occurs throughout New Jersey, Delaware, and Maryland is discussed as well as the Magothy and Monmouth formations with which it is in contact. A table giving the approximate correlation of the Atlantic Coast Cretaceous formations and their European equivalents is also given.

CLARK, WILLIAM BULLOCK, SHATTUCK, GEORGE BURBANK, and DALL, WILLIAM HEALEY.   The Miocene Deposits of Maryland.

Md. Geol. Surv., Miocene, pp. XXIII-CLV, pls. 1-9.   Balto., 1904.

Contains a full account of the Miocene strata of the State, accompanied by a map showing the distribution of the different formations.

1906.

MILLER, BENJAMIN L.   Description of the Dover Quadrangle (Delaware-Maryland-New Jersey).

U. S. Geol. Survey, Geol. Atlas of U. S., Folio No. 137, 10 pp., 1 fig., 2 maps. 1906.

The Dover quadrangle includes the greater portion of Kent County. The writer describes the physiographic features, the occurrence, character, and relations of the Cretaceous, Tertiary, and Quaternary formations, the geologic history, and the economic geology of the quadrangle.

SHATTUCK, GEORGE BURBANK.   The Pliocene and Pleistocene Deposits of Maryland.

Md. Geol. Surv., Pliocene and Pleistocene, pp. 21-137.   Plates. Baltimore, 1906.

Contains a full description of the surficial deposits of the State with many local details.

BERRY, EDWARD W.   Fossil Plants along the Chesapeake and Delaware Canal.

N. Y. Bot. Garden, Jour., Vol. VII, pp. 5-7, 1906.

CLARK, WM. BULLOCK, and MATHEWS, EDWARD B.   Report on the Physical Features of Maryland (with map).

Maryland Geol. Survey, Special Publication, Vol. VI, pt. I, Baltimore, 1906.

1907.

BERRY, EDWARD W.   New Species of Plants from the Magothy Formation.

J. H. U. Circ. n. s., No. 7, pp. 82-89, 1907.

CLARK, WM. BULLOCK.   The Classification adopted by the U. S. Geological Survey for the Cretaceous Deposits of New Jersey, Delaware, Maryland, and Virginia.

J. H. U. Circ. n. s., No. 7, pp. 89-91, 1907.

1910.

BERRY, EDWARD W.   Contributions to the Mesozoic Flora of the Atlantic Coastal Plain.   IV.   Maryland.

Torrey Bot. Club, Bull., Vol. XXXVII, pp. 10-29, 1910.

1911.

SINGEWALD, JOS. T., JR.   Report on the Iron Ores of Maryland.

Maryland Geol. Survey, Special Publication, Vol. IX, pt. III, Baltimore, 1911.

1914.

BERRY, EDWARD W.   Contributions to the Mesozoic Flora of the Atlantic Coastal Plain. X. Maryland.

Torrey Bot. Club, Bull., Vol. XLI, pp. 295-300, 1914.

1916.

CLARK, W. B., BERRY, E. W., and GARDNER, J. A.   The Upper Cretaceous Deposits of Maryland.

Maryland Geol. Survey, Upper Cretaceous, 2 vols, 1916.

1918.

CLARK, WM. BULLOCK.   The Geography of Maryland.

Maryland Geol. Survey, Special Publication, Vol. X, pt. I, Baltimore, 1918.

CLARK, WM. BULLOCK, MATHEWS, EDWARD B., and BERRY, EDWARD W.   The Surface and Underground Water Resources of Maryland, including Delaware and the District of Columbia.

Maryland Geol. Survey, Special Publication, Vol. X, pt. II, Baltimore, 1918.

THE PHYSIOGRAPHY OF KENT COUNTY

BY

BENJAMIN L. MILLER

INTRODUCTORY.

In that portion of the United States that slopes toward the Atlantic Ocean there are three physiographic provinces, each of which has certain distinguishing characteristics. These are known as the Appalachian Region, the Piedmont Plateau, and the Coastal Plain. These three provinces form bands of somewhat varying width that extend in a northeast-southwest direction roughly parallel to the shore line from New England to the Gulf of Mexico. All of these provinces are typically represented in Maryland. Garrett, Allegany, and Washington counties lie within the Appalachian Region province; Frederick, Carroll, Montgomery, Howard, and the northern and northwestern portions of Baltimore, Harford, and Cecil counties form a part of the Piedmont Plateau province; while the remaining portion of the State constitutes a part of the Coastal Plain province.

The elevations, the characteristics of the streams, and the lithologic character and structure of the rocks serve as criteria for the separation of these three provinces. In some places, however, there is such a gradation from one to another that some difficulty is encountered in drawing the exact boundary line. The Coastal Plain bordering the ocean is comparatively low and flat with few points rising more than 400 feet above sea level; the Piedmont Plateau is a higher-lying plain, some points rising to more than 1,000 feet above sea level; while the Appalachian Region, embracing the Appalachian Mountains is a much more rugged region lying at a considerably greater altitude.

The streams of the three provinces are essentially different. The tide-water estuaries of the Coastal Plain, occupying broad open valleys, form a striking contrast to the swift streams of the other two provinces which flow in steep, rock-walled gorges; while the superimposed meandering streams of the Piedmont Plateau are markedly unlike the Appalachian streams which flow in structural valleys. But probably the greatest distinction between the three provinces is due to the characters of the rocks. The unconsolidated sediments of the Coastal Plain, dipping gently toward the ocean, are sharply separated from the contorted, metamorphosed and igneous intruded strata of the Piedmont Plateau, while these in turn can be readily distinguished from the unmetamorphosed Appalachian Region limestones and sandstones that have been thrown into broad open folds, forming longitudinal ridges and valleys with a northeast-southwest trend.

Kent County is entirely within the Coastal Plain province, though the Piedmont Plateau lies only a few miles to the northwest. The adjoining counties of Cecil in Maryland and Newcastle in Delaware both contain portions of the Piedmont Plateau.

TOPOGRAPHIC DESCRIPTION

The most prominent features of the topography of Kent County are the numerous tide-water bays, creeks, and rivers that indent its shores and extend, in some cases, many miles inland.

The relief of the county is slight, there being only a little more than 100 feet difference between the lowest and highest portions of the county. From mean sea level, to which the land descends on the north, west, and south sides there is a gradual ascent to the uplands forming the stream divides, where the greatest elevations occur. As shown on the topographic map there are two areas with an elevation slightly exceeding 100 feet above sea level. One of these is located on Stillpond Neck and the other a short distance southwest of Kennedyville. The greater portion of the county forms the broad divide between the Sassafras and Chester River estuaries and a portion of the divide separating the Delaware and Chesapeake Bay drainage basins. This divide rises to a height a little more than 60 feet in the eastern part of the county and to about 80 feet in the western portion.

Within Kent County three different topographic features worthy of especial attention may be distinguished, namely, the tidal marshes, the Talbot plain, and the Wicomico plain. These vary greatly in the areas which they occupy but are principally unlike in the elevations at which each is found.

TIDAL MARSHES.

The first of these topographic features to be described consists of the tidal marshes which border the estuaries and are especially abundant in the southwestern portion of the county. They lie at a level so low that they are sometimes inundated by unusually high tides. Many of these marshes were formerly embayments from the larger estuaries or of Chesapeake Bay but in time have been so filled with material washed from the adjoining land surfaces and by the accumulation of vegetable debris that they have been converted into marshes. Many instances of marshes of this kind in process of formation can be seen at many places in the county. Small sand bars attached to one shore grow out across the mouths of these embayments until they finally meet the opposite shore. In places these barrier beaches impound considerable bodies of tidal water which, when finally filled to sea level, form extensive marshes. The accompanying illustration (Plate IX, Fig. 1) shows one of these bars which has formed across Lloyd's Creek and which in time may convert that estuary into an inland lagoon and finally into a marsh. Similar bars occur at the mouths of Churn, Worton, and Fairlee Creeks, while along the Bay shore in the vicinity of Tolchester Beach there are several lagoons that no longer have any surface connection with the waters of Chesapeake Bay.

These tide-water marshes are filled with a growth of sedges and other marsh plants, which aid in filling up the depressions by serving as obstructions to the mud carried in by small streams and by causing the accumulation of vegetable debris.

In a few places it would be possible to drain some of these marshes, as has been done in the vicinity of the Delaware River, but most of them lie too low to make drainage possible without an expenditure of money in excess of the probable returns.

TALBOT PLAIN.

The term plain is used in this discussion in a somewhat specialized sense, to include the terraces along the stream valleys and their continuations over the interstream areas, where they are true plains. The Talbot plain is defined on the geologic map as the region over which the materials constituting the Talbot formation have been spread. It borders the tidal marshes and extends from tide to an elevation of about 45 feet. This plain borders the larger streams and extends along the shore of Chesapeake Bay where it is best developed in the vicinity of Tolchester Beach and Rockhall with a width varying from 3 to 6 miles. It there exhibits its prominent characteristics of low relief and a general plain-like character. For miles there is apparently no difference in elevation whatever, the whole region being so flat that it would seem to be not well drained. In general, however, it is drained throughout; there is scarcely a marshy area in it. In places the waves of Chesapeake Bay and the estuaries have cut low cliffs from 3 to 15 feet in height in it.

The Talbot plain extends up the valleys of the Sassafras and Chester rivers, becoming gradually narrower as it reaches farther inland. It has, however, been greatly dissected by tributary streams so that it is seldom continuous for any considerable distance.

WICOMICO PLAIN.

The Wicomico plain lies at a higher level than the Talbot, from which it is in many places separated by an abrupt rise or escarp-

FIG. 1.—BAY SHORE AT WORTON POINT SHOWING BLUFF OF RARITAN
AND PLEISTOCENE MATERIALS.

FIG. 2.—BAY SHORE AT MOUTH OF LLOYD'S CREEK SHOWING MATAWAN FORMATION
WITH FERRUGINOUS NODULES.

ment varying in height from a few feet to 10 or 12 feet which is especially well developed near Melitota, Sandy Bottom, and Langford. This escarpment is often wanting, so that at some points there seems to be a gradual passage from the Talbot plain to the Wicomico. It is found, however, at so many places that there is little difficulty in determining the line of separation between the two plains. The base of the escarpment stands at an elevation of about 40 feet. From that height the Wicomico plain extends upward to an elevation of about 100 feet. At this higher elevation in adjoining regions it is separated from the next higher plain by another escarpment.

The Wicomico plain is the best developed of the three different topographic divisions within the region. It occupies the greater portion of the county and forms the broad divide between the Delaware and the Chesapeake Bay drainage systems. The Wicomico plain is very similar to the Talbot plain with the exception that it occupies a higher elevation. Along its borders it slopes very noticeably toward the lower plain, but in the interior it is exceedingly flat and monotonous. Over large areas and for distances of several miles there will not be a difference in elevation of more than 5 or 6 feet between any two of its portions. This is especially the case in the eastern half of the county. In the western part of the county through the cutting back of the small streams it is much more rolling, so that in some portions of that region its plain-like character is not preserved. On the necks of land in the northwestern portion of the county the plain has been so modified by stream erosion that a rather irregular rolling surface has been produced. In the main portion of the county the Wicomico plain is continuous and forms the divides between nearly all of the larger streams and many of the minor tributary streams. Elsewhere in Maryland a plain called the Sunderland is situated above the Wicomico and bears the same relation to the Wicomico as that plain bears to the Talbot. The

Sunderland plain extends from about 100 feet to about 180 feet above sea level.

THE DRAINAGE OF KENT COUNTY

The drainage of Kent County is comparatively simple, as a result of the simple structure of the formations and the contiguity of the region to the Delaware and Chesapeake bays. Except in a few parts all of the county is naturally drained, some areas principally by underground drainage, as is the case with the district bordering Delaware midway between the Sassafras and Chester rivers. All of the western half of the county is well drained by streams, for in that region the estuaries of the Chesapeake Bay extend inland a number of miles and the side tributaries cut back to the crests of the divides. In the eastern half, however, in the vicinity of Massey and Golts, streams are entirely absent over considerable areas, and were it not for the porous character of its soils this upland would be covered with marshes. During the rainy season water does stand on the surface, and in some places it has been necessary to dig series of ditches to connect with the natural streams. In other places, however, no ditches have been dug and there the water escapes slowly underground. The sandy surface soil is underlain by a gravel bed, so that conditions are very favorable for underground drainage.

STREAM DIVIDES.

As Kent County lies between Chesapeake Bay and Delaware Bay, both of which are at sea level, it would naturally be expected that the watershed between the two drainage systems would divide the peninsula into two symmetrical parts; yet, notwithstanding the fact that there is little in the character of the materials, the positions of the beds, or the comparative proximity to tide water to cause the streams emptying into Chesapeake Bay to cut more rapidly than those emptying into Delaware Bay, the water parting is considerably nearer Delaware Bay and the entire drainage of Kent County passes into Chesapeake Bay.

The asymmetrical character of the divide is much more pronounced in areas farther south, where the streams tributary to Chesapeake Bay extend to within a few miles of the ocean. The cause of this asymmetry is believed to date back to a period when the whole region stood at a higher level, when Susquehanna River emptied into the ocean a considerable distance east of Cape Henry and Cape Charles, when the mouth of Delaware River lay east of Cape Henlopen and Cape May, and when the peninsula was much wider than it is now, comprising land on both sides now submerged by the waters of Delaware River, Delaware Bay, the Atlantic Ocean, and Chesapeake Bay. The old channels of Susquehanna and Delaware rivers can still be traced throughout a great portion of Chesapeake and Delaware bays, notwithstanding the fact that recent deposition has in many places obliterated the depressions. An examination of the soundings in the two regions indicates a deeper channel in Chesapeake Bay, and presumably before the recent submergence of this region the waters of the lower course of the Susquehanna flowed in a channel considerably lower than that occupied by the waters of the lower course of the Delaware. This permitted the streams tributary to the Susquehanna to extend their headwaters much more rapidly than the Delaware River tributaries and thus gradually shifted the divide to the eastern portion of the peninsula.

TIDE-WATER ESTUARIES.

The lower courses of almost all the larger and many of the smaller streams emptying into Chesapeake Bay have been converted into estuaries through submergence which has permitted tide-water to pass up the former valleys of the streams. In the early development of the country these estuaries were of great value, since they are navigable for several miles from their mouths and thus afforded the means of ready transport of the produce of the peninsula to market. Even the advent of railroads has not rendered them value- less, for much grain and fruit are still shipped to market on steam- ers and small sailing vessels which pass many miles up these estuaries. Steamboats from Baltimore pass up Sassafras River as far as Fredericktown, while freight sailing vessels go within a short distance of the town of Sassafras. Chester River is similarly navi- gable almost to the town of Millington. The estuaries also furnish good fishing grounds and during certain seasons are frequented by wild water fowl in such numbers that Chesapeake Bay and its tributaries have long been known to sportsmen as among the finest hunting grounds in the country.

The Sassafras and Chester rivers are the most important estu- aries of the county though many of the smaller estuaries are navi- gable for a distance of several miles from their mouths. Among these are Worton and Grays creeks and Langford Bay. Sassafras River is the deepest of these estuaries. The maximum depression in this stream lies just west of Ordinary Point, where recent charts of the Coast and Geodetic Survey show 50 feet of water. Another depression near Cassidy wharf, has a depth of 48 feet. Exclusive of these deep places, the channel as far up as Fredericktown has an average depth of about 14 feet. Beyond this point its depth gradu- ally decreases. Because of the channel of the Sassafras River estuary being so deep it has been investigated frequently by Federal commissions appointed to examine and report upon a waterway to connect the waters of Chesapeake and Delaware bays. In Chester River there is a dredged channel 8 feet deep from Spry Landing to Crumpton, and 6 feet deep from Crumpton to the mouth of Mills Branch.

Sassafras River and its tributary estuaries are bordered by nearly vertical bluffs 10 to 60 feet in height, or by slopes which rise rapidly to the height of the broad upland within the distance of half a mile from the river. That the present estuaries have not carved the bluffs that border them is very evident, since they are now doing little erosive work themselves. The small waves that are produced at times of strong westerly winds are the only notable agents of erosion. Such waves are frequently able to remove the finer debris that accumulates as talus at the foot of the cliffs, especially in the early spring, but are not strong enough to do much undercutting. The present cliffs represent bluffs that bordered the valleys of streams whose flood plains as well as channels are now covered by the estuarine waters.

The water in the estuaries is fresh or very slightly brackish and ebbs and flows with the tide. There is seldom any distinct current to be noticed and such as is seen is due to the incoming or outgoing tide and appears to be nearly as strong when moving upstream as when moving in the opposite direction.

At Turkey Point, the southern extremity of Elk Neck, in Cecil County, the average height of the tides above mean low water is 2 feet.

MINOR STREAMS.

Besides the estuaries which form so prominent a feature of the county there are numerous minor streams which drain into these estuaries. At the head of each estuary there is a small stream which, in almost every case, is very much shorter than the estuary itself. Some of the estuaries, particularly that of Sassafras River, continue as such almost to the sources of the tributary streams. Further, those streams which flow into the estuaries from the side are seldom more than a few miles in length.

Although nearly the entire region lies less than 100 feet above the sea, and although these minor streams descend gradually from the divides to sea level, yet they furnish considerable water power. This is utilized by numerous mills that are located on various streams which empty into the estuaries. The map shows these numerous millponds and also indicates their relatively large size.

Because of the gentle slope of the stream channels, a dam of ordinary height may form a pond that extends for a mile or more up the stream.

An inspection of the county topographic map shows that the tributary streams of Chester River present different characteristics than those flowing into the Sassafras River. The former have broad valleys with gentle slopes while the latter are more numerous and flow in deep narrow valleys. These differences are accounted for partially by the dip of the strata which is toward the southeast but mainly by the fact that the Cretaceous strata which outcrop along the Sassafras River are worn away much more readily than are the Tertiary strata that outcrop along the Chester River. Still another reason is that the waves can do more effective work in Sassafras River because of its greater depth and because it is more exposed to the northwest winds. All along the eastern shore of Chesapeake Bay there are evidences of much greater erosion being accomplished by the strong northwest winds of the winter season than by winds blowing from any other quarter.

TOPOGRAPHIC HISTORY

The history of the development of the topography as it exists to-day is not complicated. The topographic features were formed at several different periods, during all of which the conditions must have been very similar. The physiographic record is merely the history of the development of the two plains already described as occupying different levels, and of the present drainage channels. The plains of Kent County are primarily plains of deposition which, since their formation, have been more or less modified by the agencies of erosion. Their deposition and subsequent elevation to the heights at which they are now found indicate merely successive periods of depression and uplift. The drainage channels have throughout most of their courses undergone many changes; periods of cutting have been followed by periods of filling, and the present valleys and basins are the results of these opposing forces.

THE WICOMICO STAGE.

When the Coastal Plain had been above water for a considerable time after the close of the Sunderland deposition a gradual submergence again occurred, so that the ocean waters once more encroached on the land. This submergence seems to have been about equal in amount through a large portion of the district, showing that the downward movement was without deformation. The sea did not advance upon the land as far as it did during the previous submergence. At many places along the shore the waves cut cliffs into the deposits that had been laid down during the preceding epoch of deposition. Throughout many portions of the Coastal Plain at the present time these old sea cliffs are still preserved as escarpments, ranging from 10 to 15 feet in height. Where the waves were not sufficiently strong to enable them to cut cliffs it is somewhat difficult to locate the old shore line. During this time all of Kent County was submerged. The Sunderland deposits were largely destroyed by the advancing waves and redeposited over the floor of the Wicomico sea.

Although the Wicomico submergence permitted the silting up of the submerged stream channels, yet the deposits were not thick enough to fill them entirely. Accordingly, in the uplift following Wicomico deposition the large streams reoccupied their former channels, with perhaps only slight changes. New streams were also developed and the Wicomico plain was more or less dissected along the water courses, the divides being at the same time gradually narrowed. This erosion period was interrupted by the Talbot submergence, which carried part of the land beneath the sea and again drowned the lower courses of the streams.

THE TALBOT STAGE.

The Talbot deposition did not take place over so extensive an area as was covered by that of the Wicomico. It was confined to the old valleys and to the low stream divides, where the advancing

waves destroyed the Wicomico deposits. The sea cliffs were pushed
back as long as the waves advanced, and now stand as an escarpment that marks the boundaries of the Talbot sea and estuaries.
This is the Talbot-Wicomico escarpment, previously described. At
some places in the old stream channels the deposits were so thick
that the streams in the succeeding period of elevation and erosion
found it easier to excavate new courses than to follow the old ones.
Generally, however, the streams reoccupied their former channels
and renewed the corrosive work which had been interrupted by the
Talbot submergence. As a result of this erosion the Talbot plain
is now in many places somewhat uneven, yet it is more regular than
the Wicomico plain which has been subjected to denudation for a
longer period.

THE RECENT STAGE.

The land probably did not long remain stationary with respect
to sea level before another downward movement began. This last
subsidence is probably still in progress. Whether this movement
will continue much longer cannot, of course, be determined, but
with respect to Delaware River there is sufficient evidence to show
that it has been in progress within very recent time and undoubtedly
still continues. Many square miles that had been land before this
subsidence commenced are now beneath the waters of Chesapeake
Bay and its estuaries, and are receiving deposits of mud and sand
from the adjoining land.

FIG. 1.—VIEW SHOWING CROSS BEDDING IN THE WICOMICO FORMATION NEAR BETTERTON.

THE GEOLOGY OF KENT COUNTY

BY

BENJAMIN L. MILLER

INTRODUCTORY.

The geologic formations represented in Kent County range in age from Cretaceous to Recent. Deposition has not been continuous, yet none of the larger geologic divisions since Cretaceous time is entirely unrepresented. Periods when deposition occurred over part or the whole of the region are separated by other periods, of greater or less duration, in which the entire region was above water and erosion was active. The deposits of all the periods, except those of the Pleistocene, are similar in many respects. With a general northeast-southwest strike and southeast dip, each formation disappears southeastward by passing under the next later one. In general also the shore during each successive submergence evidently lay a short distance southeast of the line it occupied during the previous submergence. There are a few exceptions to this, however, which will be noted in the descriptions that follow. Thus, in passing from the northwest to the southeast one crosses successively the outcrops of the formations in the order of their deposition.

TABLE OF GEOLOGIC FORMATIONS.

System	Series	Group	Formation
Quaternary-Pleistocene............................Columbia			{ Talbot { Wicomico
Tertiary................ {	Miocene...............Chesapeake........		Calvert
	Eocene...............Pamunkey.........		Aquia
Cretaceous .... {	Upper Cretaceous		{ Monmouth { Matawan } Magothy { Raritan
	Lower Cretaceous.......Potomac................		

THE CRETACEOUS SYSTEM

LOWER CRETACEOUS.

THE POTOMAC GROUP.

The Potomac group of the Coastal Plain consists of highly colored gravels, sands, and clays which outcrop along a sinuous line that extends from Delaware to Virginia, passing near the cities of Philadelphia, Wilmington, Baltimore, and Washington. The Potomac deposits are of great value because of the excellent brick clays which they contain. Of the three formations that have been recognized as composing the Potomac group in Maryland, the Patapsco, the Patuxent, and the Arundel, none is represented within the county. The Patapsco and Patuxent formations outcrop a short distance to the northwest of Kent County and probably underlie this entire county though they do not appear at the surface.

UPPER CRETACEOUS.

THE RARITAN FORMATION.

The formation receives its name from Raritan River, New Jersey, in the basin of which it is typically developed. The name in its present usage was proposed by W. B. Clark in 1892 (Ann. Rept. Geol. Survey N. J. for 1892-93, pp. 169-243), although the term had been loosely applied to these deposits by earlier writers. It includes the deposits long called the Plastic or Amboy clays by the New Jersey Geological Survey.

Areal Distribution.

In its wider distribution the Raritan formation has been traced from Raritan Bay, New Jersey, to the basin of the Potomac River. In Kent County the outcrops of the Raritan are entirely confined to the extreme northwest corner of the county where almost every bluff along the Bay and creeks, between the mouths of the Sassafras

River and Worton Creek contains exposures of this horizon. Good sections can be seen about one-half mile below Harris Wharf, at Kinnairds Point, Rocky Point, and Worton Point.

Since the Raritan dips to the southeast it.seems probable that it underlies the entire county. At Middletown, Delaware, about 6 miles northeast of the northeast corner of the county, the formation was reached in an artesian well at a depth of 425 feet.

Character of Materials.

The materials of the Raritan are extremely variable in character. Variegated clays, horizontally stratified, and cross-bedded sands and gravels, and occasional ledges of sandstones and conglomerates are all represented within the formation.

The character of its materials changes at many places very abruptly, both horizontally and vertically. Iron in some form, chiefly as an oxide, is commonly present and forms the cementing material for the locally indurated layers of sandstones and conglomerates. The loose sands interbedded with impervious plastic clays form important water-bearing beds and in several places furnish artesian water, as described later.

The following sections illustrate the general character of the Raritan formation in this county.

CLIFF SECTION, 1 MILE EAST OF HOWELL POINT.

		Feet
Wicomico.	Stratified sands, clays, and gravels, coarsest materials at base	30
Raritan.	Very fine white to light-drab colored sand containing small flakes of mica	10
	Coarse red sand, poorly exposed because of cliff talus perhaps in thickness as much as	10
	Total	50

SECTION ½ MILE BELOW HARRIS WHARF.

		Feet
Wicomico.	Cross-bedded sands and gravels...............	18
Magothy.	Laminated black clay containing fragments of white sand	5
Raritan.	Loose buff to iron-yellow stratified sands, exposed	16
	Total	39

SECTION 1½ MILES BELOW HARRIS WHARF.

		Feet
Wicomico loam.	....	6
	Variegated pink and yellow clay..............	5
	Gravel band	3
	Yellow clay	½
	Gravel band	2½
	Variegated clay pink and light green..........	3
	Coarse gravel and sand.....................	12
Raritan.	Very coarse light-yellow sand	4
	Indurated ferruginous sand	1
	Yellow sand	4
	Pink clay	½
	Pure white sand	1
	Concealed by detritus to water's edge..........	10
	Total	51

SECTION AT ROCKY POINT.

		Feet
Talbot.	Stratified loam, sand, and gravel exposed in near-by area.	
Raritan.	Stratified coarse red sandstone firmly indurated by iron oxide (see illustration), exposed to water's edge	12

SECTION AT WORTON POINT.

		Feet
Talbot.	Loam	6–8
	Coarse gravel	8
Raritan.	Pink to red fine textured compact plastic clay, exposed to water's edge	12
	Total	26–34

Paleontologic Character.

The fossils of the Raritan formation consist largely of plant remains which have been recognized at many different localities in New Jersey and Maryland. The known flora of the formation includes ferns, conifers, cycads, monocotyledons, and dicotyledons. There is a wide range of genera and species, especially of the dicotyledons, many of which belong to living genera. The known fauna is very limited, consisting of a few pelecypods, a plesiosaurian bone, and possibly an insect.

Strike, Dip, and Thickness.

The thickness of the Raritan formation at its outcrop in Kent County, where it has been subjected to excessive erosion, does not exceed 40 feet at any point. Elsewhere in Maryland where the contact with the next younger formation is shown, the thickness is over 200 feet. It thickens gradually southeastward, down the dip. The author believes that at least 500 feet of the materials penetrated by the Middletown, Delaware, artesian well should be referred to this formation. The strike is northeast and southwest and the dip is about 30 feet to the mile.

Stratigraphic Relations.

The Raritan overlies the Patapsco formation, where the lower contact has been observed, with which it is unconformable. It is separated from the overlying Magothy deposits by another marked unconformity. In the region of its outcrop Pleistocene deposits of the Talbot, Wicomico, and Sunderland formations overlie the edges of the formation and generally conceal the deposits from view except where erosion has removed these later beds.

THE MAGOTHY FORMATION.

The Magothy formation takes its name from the excellent exposures of the beds of this age along the Magothy River in Anne Arundel County.

Areal Distribution.

The Magothy formation outcrops in the extreme northwestern portion of Kent County in a narrow band 2 to 3 miles in width that extends from Betterton to Worton Creek. Over the divides it is concealed by the loam, sands, and gravels of the Pleistocene, thus limiting its exposures to the cliffs cut by the waters of Chesapeake Bay and tributary streams. The best exposure in the county occurs at Betterton, while other good sections can be seen near Harris Wharf, on the south shore of Stillpond Creek and at Worton Point.

Character of Materials.

The Magothy formation is composed of extremely varied materials and may change abruptly in character both horizontally and vertically. Loose sands of light color are the most prominent constituents. These sands usually show fine laminations and locally considerably cross-bedding. The sand consists of coarse, rounded to subangular quartz grains which vary in color from pure white to a dark ferruginous brown. At many places lenses or bands of brown sand occur within the lighter colored sands. While normally the deposits of sand are loose, yet locally the iron derived from this and adjacent formations has firmly cemented the grains together to form an indurated iron sandstone or conglomerate. Small pebbles are apt to be present near the base of the deposits.

The argillaceous phase of the Magothy is very prominent in some localities, although it is usually subsidiary to the arenaceous phase. The clay commonly occurs in the form of small pellets in the sand or as fine laminae alternating with the sand layers. Drab is the characteristic color of the Magothy clay, but occasionally the presence of considerable vegetable remains renders it black. The vegetable material may be finely divided or may occur in the form of large pieces of lignite. The lignite is in many places impregnated with pyrite and marcasite which are also found associated with the lignite in the form of oblong to spherical concretions several inches in diameter.

SECTION ½ MILE WEST OF BETTERTON WHARF.

		Feet
Wicomico.	Loam	4
	Coarse red argillaceous sand	3
	Coarse gravel containing limonite concretions..	4
	Coarse red sand	6
	Gravel band	1
	Light-colored clay	½
	Light-colored sand	1
	Dark-colored sand	¼
	Light-colored sand	½
	Very coarse light yellow cross-bedded sand containing many solitary pebbles and lenses of gravel	18
	Coarse gravel	2
	Very coarse light-yellow cross-bedded sand containing thin gravel bands	13
	Coarse gravel	¼
	Very coarse pebbly light-yellow sand	6
	Coarse gravel	½
	Coarse gravelly light-yellow sand	9
	Ironstone conglomerate	¼
Magothy.	Black sandy clay filled with lignite. In the larger fragments of lignite there is considerable pyrite and marcasite which impregnates the lignite or forms nodular concretions about the stems. Exposed to water's edge	19
	Total	88¼

At Betterton the Magothy is only about 9 feet above sea level while it disappears beneath tide water in less than ½ mile east of the Betterton wharf. Westward from the locality where the above section was taken the lignite and pyrite become less abundant in the Magothy while the sandy clay becomes light-drab in color. The flakes of mica likewise gradually become more numerous. In one place a thin lens of small pea gravel composed of white vein-quartz was observed near the top of the Magothy. The strata rise to the westward and at a point 1½ miles east of Howell's Point they disappear by erosion. From that point westward the Raritan outcrops at the base of the cliffs beneath the mantle of Pleistocene materials.

At Worton Point the Magothy is well exposed though it is only a few feet thick. It there overlies the Raritan unconformably and the contact between the two formations can be readily seen to descend from about 10 feet above sea level to tide in a distance of about ⅛ mile, and this descent is approximately parallel to the strike of the formations. At this place the Magothy contains much lignite, marcasite, and pyrite. Beautiful concretions of pyrite and marcasite can be picked up along the beach or dug out of the plastic clay.

The arenaceous phases of the Magothy, which elsewhere are most common, are exhibited in several exposures in the vicinity of Stillpond Creek. In a bluff just above the mouth of Stillpond Creek there is an exposure of about 6 feet of Magothy materials. These consist of drab and dark-colored clays containing pyrite and marcasite interbedded with finely laminated buff to white sands. On the south side of Stillpond Creek there is an exposure of about 12 feet of light-green and yellow-brown mottled sand, which in a few places contains small pockets of glauconitic sand.

Paleontologic Character.

No organic remains have thus far been recognized in the Magothy formation, in this county, although a considerable flora has been described from it at Grove Point in Cecil County.

Strike, Dip, and Thickness.

In Kent County the Magothy formation is less than 40 feet thick, but in its wider extent its thickness is extremely variable, reaching a maximum of about 100 feet. This variability is due to greater deposition in some regions than in others and also to the removal of considerable material in certain areas. It dips southeastward at about 30 feet to the mile and disappears at tide level near the mouths of the estuaries tributary to Chesapeake Bay in the northwestern corner of the county and does not again appear

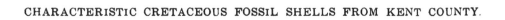

1

2

3

4

5

CHARACTERISTIC CRETACEOUS FOSSIL SHELLS FROM KENT COUNTY.

at the surface in this region. In all probability it underlies the remainder of the county and should be recognized in detailed deep-well sections in the central and eastern parts. The strike is roughly parallel to that of the other Coastal Plain formations—from northeast to southwest.

Stratigraphic Relations.

The Magothy formation is included between the Raritan and Matawan formations and is separated from each by an unconformity. The line of contact between the Magothy and the Raritan is very irregular, indicating a considerable erosion interval between the times of their deposition. In many places the Magothy deposits fill pockets and old channels in the Raritan. The unconformity between the Magothy and the Matawan is not so plainly marked.

THE MATAWAN FORMATION.

The Matawan formation received its name from Matawan Creek, a tributary of Raritan Bay, in the vicinity of which the deposits of this horizon are typically developed. The name was proposed by Wm. Bullock Clark in 1894 (Jour. Geol., vol. 2, pp. 161-177) and replaced the term Clay Marls, previously used by the New Jersey geologists.

Areal Distribution.

In Kent County it is rather poorly developed at the surface, appearing only in places where stream erosion has removed the overlying Pleistocene. Its outcrops are confined to the northwestern portion of the county in a narrow belt, 2 to 3 miles in width, extending from Lloyd Creek southwest to Fairlee Creek. The best exposures occur in the steep bluffs about Lloyd Creek, especially near its mouth, and at the head of Stillpond Creek estuary. It undoubtedly underlies all that part of the county that is southeast of its line of outcrop. In its broader distribution throughout the Coastal Plain the Matawan formation outcrops as a continuous series of deposits from Raritan Bay to Potomac River.

Character of Materials.

The Matawan consists chiefly of glauconitic sand intimately mixed with dark-colored clay, all quite micaceous. In some places the deposits consist almost entirely of black clay; in others, particularly where the upper beds are exposed, the arenaceous phase is predominant and the beds may consist entirely of sands that vary in color from white to dark greenish-black. When the glauconite in the beds is decomposed the iron oxidizes and the materials are stained reddish brown and may even become firmly indurated by the iron oxide. Iron pyrite is locally a common constituent and a small layer of gravel is sometimes found at the base of the formation. The character of the formation as developed in this county is shown in the following section.

SECTION AT MOUTH OF LLOYD CREEK
2½ Miles East of Betterton.

		Feet
Wicomico.	Numerous gravels, boulders, fragments of ferruginous conglomerate in matrix of loose white to yellow sand	10–12
Monmouth.	Rich, brownish-yellow sand, containing numerous exceedingly irregular ironstone concretions roughly arranged in layers. In certain places the sandy matrix is gray	20
Matawan.	Mottled dove-colored to brown sand in places dark and light sands mixed resembling pepper and salt, containing small pebbles about the size of a pea in the upper portion. In the lower 3½ feet exposed there are numerous oblong concretions consisting of very hard brown to black sandstone, many having the shape of an hour-glass. In size they range from 1 to 4½ feet in height, about 1½ feet thick, and 1½ to 4 feet in width. Exposed to water's edge	28
	Total	58–60

A concretion obtained at this locality is on exhibition at the U. S. National Museum in Washington.

About 1 mile northeast of Stillpond, where the road crosses a small stream, there is a good exposure of Matawan dark micaceous sand, while just below the mill dam on Stillpond Creek there is another good exposure. In the latter locality the changes which take place on weathering are well shown. The line between the weathered and unweathered portion is so distinct that it suggests a stratigraphic break. The upper part is yellowish-brown to gray in color and contains many indurated bands of ironstone, resulting from the segregation of iron oxide formed during the decomposition of the glauconite, while the lower unweathered portion is a dark-colored compact micaceous argillaceous sand.

Paleontologic Character.

The Matawan formation has yielded few fossils in Kent County. At the milldam on Stillpond Creek a few Exogyra shells and a fragment of a crab's claw were found. In New Jersey, and elsewhere in Maryland the formation has yielded a varied fauna of foraminifera, pelecypods, gastropods, scaphopods, and ammonites.

Strike, Dip, and Thickness.

In its northern extension the formation has a thickness of about 220 feet, but it thins to the south and in the vicinity of Potomac River is only 20 feet thick. At its outcrop in Kent County it is about 40 to 50 feet in thickness. Like many other Coastal Plain formations, the beds thicken as they dip beneath later deposits, but the records of wells which have penetrated the formation east of line where it disappears from view are too general to permit a determination of the amount of thickening. The strike and dip do not differ from those of the preceding formation.

Stratigraphic Relations.

An unconformity separates the Matawan from the underlying Magothy formation, but the Matawan is conformably overlain by the Monmouth. The separation between the Matawan and Monmouth is made chiefly on the basis of change in lithologic character, but in part on fossil content. Although some organic forms range through both the Matawan and Monmouth, yet each formation has a few characteristic ones, the assemblages in each being on the whole quite distinctive.

THE MONMOUTH FORMATION.

The name Monmouth was first proposed by Wm. Bullock Clark in 1897 (Bull. Geol. Soc. Amer., vol. 8, pp. 315-358) when it was decided to combine in a single formation the deposits formerly included in the Navesink and Redbank formations. This name was suggested by Monmouth County, N. J., where the deposits of this horizon are characteristically developed. It was employed for the term Lower Marl Bed of the earlier workers in New Jersey.

Areal Distribution.

The Monmouth formation outcrops along the stream valleys in a belt about 3 to 5 miles broad, which extends across the northwestern portion of the county in a northeast-southwest direction. Over the divides the Monmouth deposits are concealed from view by the materials of the Wicomico formation, while near the streams they are often covered by the Talbot loam. Only where the streams have been able to remove this capping of younger materials is the Monmouth formation exposed to view. The rather deep valleys, with their precipitous bluffs, along Sassafras River and its tributaries, Turner, Freeman, Island, and Mill creeks, afford many excellent exposures. To the southwest, it is best exposed in the escarpment between the Talbot and Wicomico formations in the vicinity of Melitota and Fairlee while it is also exposed in the headwaters of some small streams. In its wider distribution the formation has been recognized by outcrops in a zone extending from Atlantic Highlands to a point a short distance beyond Patuxent River.

Character of Materials.

The formation is prevailingly arenaceous in character and unconsolidated except where locally indurated by the segregation of ferruginous material derived from the glauconite. The sands composing the Monmouth deposits vary in color from reddish-brown to dark green or nearly black. The fresh material always contains considerable glauconite and this gives to the deposits their dark color. In their more weathered portions the sands generally range in color from rich brown to reddish-brown, but at some places they are dark gray.

The Monmouth deposits of New Jersey, which are continuous with those of this region, have been divided into three members. These divisions have not been recognized in Kent County.

The lower beds of the latter area are somewhat more glauconitic than the upper but are not sharply separated from them. In the vicinity of Cassidy wharf, on the Cecil County side of the Sassafras River, the lower marly beds have a thickness of about 20 feet, while near Ordinary Point they are about 45 feet thick. The material consists of fine, slightly micaceous sand so intermixed with brown iron-stained sand as to give the whole a mottled appearance. Within the iron-stained portions are found pockets of gray-green glauconitic sand. Under the microscope it is seen that the grains of sand from the more ferruginous parts are completely coated with iron, while those from the lighter colored pockets are entirely free from it. On the south side of Sassafras River, near Turner Creek wharf, there is an exposure of about 40 feet of lower Monmouth materials. In the lower portion of the section numerous iron crusts and concretions are present in a brown sand. Most of the iron concretions are exceedingly irregular, but some are pipe-like, long, and straight, and usually hollow. A few iron-incrusted fossil casts are present in this part of the section. The upper 20 feet is composed of light-colored glauconitic sand containing some soft lime concretions and a few fossil casts. A few iron crusts are also present.

The lower marly beds of the Monmouth occur also at a few places along the tributaries of Bohemia Creek, Cecil County. On the north side of the creek, just east of the bridge, marl for fertilizing purposes has been dug at several places, but none of these marl pits are now worked.

The upper portion of the Monmouth formation in this region consists of beds of rather coarse sands which at some places are decidedly red in color, although usually a reddish-brown. Here and there in this portion of the formation are pockets containing considerable glauconitic sand. The sand is frequently casehardened and occasionally firmly cemented by ferruginous material. The sands are exposed at many places along Sassafras River. In the headwaters of some small streams near Locust Grove the Monmouth appears as a bright green glauconitic sand, while near Melitota nearly all the glauconite has been decomposed, leaving a gray to yellowish-brown sand.

Paleontologic Character.

The Monmouth formation is generally very fossiliferous and the forms are usually well preserved. They consist of foraminifera, pelecypods, gasteropods, and cephalopods. Among the most abundant fossils found in the Monmouth in this area are *Exogyra costata* Say, *Gryphaea vesicularis* Lamarck, *Cucullaea vulgaris* Morton, *Cardium Kümmeli* Weller, and *Belemnitella americana* Morton. They are typical Upper Cretaceous species. Several of these are shown on the accompanying plate.

Strike, Dip, and Thickness.

The total thickness of the Monmouth formation along its out-crop in Kent County is about 80 feet. In northern New Jersey it is about 200 feet thick, but from there it steadily decreases in thickness along the strike, southwestward, until, in the valley of Patuxent River, the beds are only 10 feet thick.

Stratigraphic Relations.

The formation is conformable with the underlying Matawan and with the Rancocas which overlies it in Delaware and New Jersey. Along the Sassafras River and elsewhere in Kent County it is unconformably overlain by Eocene deposits, the Rancocas being absent from this county. Pleistocene materials conceal it from view over the divides and at some places even in the stream valleys.

THE TERTIARY

THE EOCENE FORMATIONS.

The Pamunkey Group.

THE AQUIA FORMATION.*

Areal Distribution.

The Aquia is the only Eocene formation in Kent County. Its outcrops are found along Sassafras and Chester Rivers and their tributaries, in a belt from 5-6 miles wide, that extends from the headwaters of the Sassafras River near the Delaware line to Langford Bay. Yet notwithstanding its great areal extent there are comparatively few good exposures of more than a few feet of Eocene materials. The best ones occur along the Sassafras River, in the vicinity of Georgetown and Wilson Point wharf and along the Chester River about 2 ½ miles below Chestertown. Minor exposures occur along the streams in the vicinity of Morgnec, Bigwoods, and Chestertown, and at the base of the Talbot-Wicomico escarpment near Langford and Sandy Bottom. The Aquia forma-

* Md. Geol. Survey, Eocene, 1901.

tion has been recognized in a series of disconnected outcrops that extend from a point near the border of Delaware southward through Virginia.

Character of Materials.

This formation consists usually of loose sand in which there is a considerable admixture of glauconite, the latter in places making up the body of the formation. Where the material is fresh the deposits range in color from a light blue to a very dark green, but in regions where the beds have been exposed to weathering for a considerable time they have assumed a reddish-brown to light-gray color. The beds are in most places unconsolidated, although locally some have become very firmly indurated by oxide of iron.

SECTION AT WILSON POINT WHARF ON THE SASSAFRAS RIVER.

		Feet
Talbot.	Coarse brown sand, very compact, containing isolated gravels and small gravel lenses with a quite persistent gravel band 9 to 13 feet thick at base	11
Aquia.	Green glauconitic sand, upper part intensely green and lower 5 to 8 feet lighter in color and more or less consolidated	27
	Total	38

A short distance below the locality where the above section was taken there is a firmly indurated ledge of weathered green sand, rich-brown in color, about 8 feet in thickness in which there are numerous casts of fossils.

SECTION NEAR MILLDAM NORTHEAST OF GALENA.

		Feet
Wicomico.	Gravel in a matrix of coarse sand	4
Aquia.	Brownish-yellow, very compact, weathered green-sand, grading downward into fresher material, somewhat gray in color	14
	Dark-green, coarse glauconitic sand, filled with numerous iron concretions, usually irregular in shape, although sometimes showing a slight tendency to nodular structure	10
	Total	28

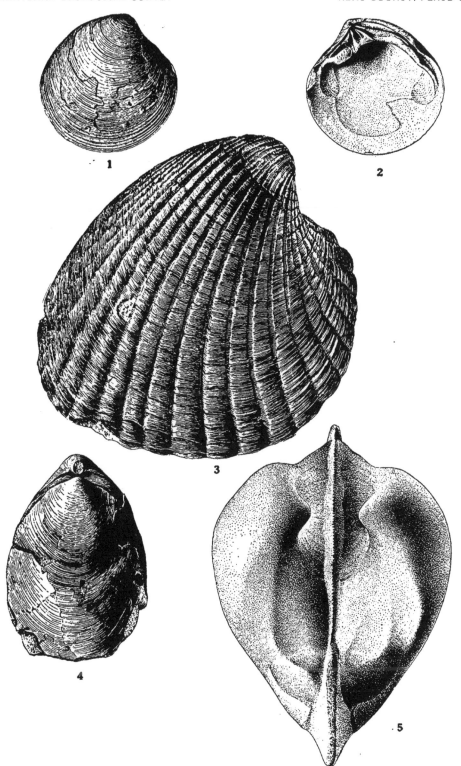

1

2

3

4

5

The best exposure of the Aquia in Kent County shows the indurated glauconitic sand which outcrops along the Chester River below Chestertown.

SECTION ON RIGHT BANK OF CHESTER RIVER
1 MILE NORTHWEST OF ROLPHS.

		Feet
Talbot.	Sand and loam	3–5
Aquia.	Very coarse indurated glauconitic sand, much oxidized and iron-stained, with abundant angular quartz pebbles, some of which are nearly ½ inch in diameter. Abundant casts of fossils including *Turritella mortoni, Panopea elongata, Protocardia lenis, Venericardia planicosta*, var.. *regia, Crassatellites alaeformis, Glycimeris idoneus, Cucullaea gigantea*	4–6
	Yellowish-red slightly indurated sand bearing a few fossil casts	5–6
	Oxidized glauconitic sand, with occasional tubes of *Vermetus.* Exposed to water's edge	4
	Total	12–16

Near Sandy Bottom the Aquia consists almost entirely of glauconitic sand, and there marl has been dug for fertilizing purposes.

Paleontologic Character.

A great many fossils are seen in the outcrops of the Aquia along Sassafras River from Georgetown to Sassafras, but most of them are poorly preserved and only a few can be identified. At Fredericktown the following forms have been recognized: *Dosiniopsis lenticularis* (Rogers), *Venericardia planicosta* (Conrad), *Cuculœa gigantea* (Conrad), and *Terebratula marylandica* Roberts. These are shown on the accompanying plate.

Strike, Dip, and Thickness.

The strike and dip correspond in general with the previously-described formations. The thickness of the Aquia exposures in Kent County is about 35 feet. Toward the south the formation thickens, reaching a total thickness of about 100 feet on the west side of Chesapeake Bay in southern Maryland.

Stratigraphic Relations.

By the transgression of the Aquia sea, the beds of this formation, which should normally overlie merely the Rancocas, have been brought into direct contact with Monmouth deposits along Sassafras and Chester rivers. On the western side of Chesapeake Bay in southern Maryland a higher member of the Eocene, the Nanjemoy formation, is exposed. The Nanjemoy is not represented in Kent County, its absence being due likewise, no doubt, to the overlap of the Calvert formation. In the divide between Sassafras and Chester rivers the Aquia is unconformably overlain by the Wicomico formation and in the valleys of these two rivers by Talbot materials, while in the area bordering the Chester River in the southeastern portion of the county it is covered unconformably by the Calvert formation.

THE MIOCENE FORMATIONS.

The Chesapeake Group.

THE CALVERT FORMATION.

The formation receives its name from Calvert County where in the well-known Calvert Cliffs bordering Chesapeake Bay its typical character is well shown.

Areal Distribution.

The Calvert, the only Miocene formation in Kent County, crops out along the Chester River and some of its tributaries in the southeastern portion of the county. In that section there are no high bluffs, the Pleistocene covering is thicker than in the western part of the county, and the slopes of the stream valleys are gentle, consequently there are few exposures of Calvert materials. The best sections observed occur on the river road about ½ mile west of Millington and at the crossing of Mills Branch on the Millington-Chestertown road. To the south and southeast of Kent County the

Calvert formation is well developed. In its wider distribution it has been recognized in New Jersey, whence it extends southward through Delaware, Maryland, and Virginia into North Carolina.

Character of Materials.

In Kent County the Calvert consists of very fine buff to white quartz sands which, in places, are streaked with alternating bands or blotches of white and buff. Diatomaceous earth, which forms such a prominent constituent of the formation elsewhere in Maryland and in New Jersey, Delaware, and Virginia, has not been observed in this county, though it is exposed in Queen Anne's County, a short distance south of Chester River. Blue sandy clay is also present in other places in Maryland and in Delaware but is unrepresented in this county.

SECTION ONE-HALF MILE WEST OF MILLINGTON.

		Feet
Talbot.	Coarse gravel in matrix of loose gray or slightly indurated yellowish-brown ferruginous sand..	7
Calvert.	Fine gray quartz sand with yellow streaks and blotches running through it, exposed.........	1½
	Total	8½

Paleontologic Character.

No fossils have been found in the Calvert in this county though elsewhere the formation contains a varied and extensive marine fauna and flora. The diatomaceous earth has yielded a great quantity of diatoms while the shell layers, developed in certain places, are composed of quantities of the remains of molluscan and other invertebrate forms of life with occasional vertebrate bones and teeth. The fossils are allied to forms now living in lower latitudes, this fact indicating a somewhat warmer climate in this region during the period of deposition of the Calvert materials. The fossils of this formation have been fully described and illustrated in the volume on the Miocene issued by the Maryland Geological

Survey. The accompanying plate shows the more characteristic fossil shells, any of which may be found in the southeastern part of the county.

Strike, Dip, and Thickness.

The thickness of the Calvert in Kent County is probably not more than 15 feet. South of this region it gradually thickens as it passes beneath strata of later age. At Crisfield a well section indicates over 300 feet of Calvert materials.

Stratigraphic Relations.

In this region the Calvert unconformably overlies the Aquia formation while it is in turn overlain by deposits of the Talbot and Wicomico formations between which there are likewise marked stratigraphic breaks. On the western shore of Maryland the Calvert overlies the Nanjemoy formation of the Eocene and is overlain by Miocene strata belonging to the Choptank formation.

THE PLEISTOCENE FORMATIONS.

The Columbia Group.

The Pleistocene formations of the Atlantic Coastal Plain are united under the name Columbia group. They have many characteristics in common, due to their similar origin. They consist of gravels, sands, and loam, which are stratigraphically younger than the Brandywine or Bryn Mawr formation. The Columbia group has been divided in Maryland into three formations: the Sunderland, Wicomico, and Talbot, the last two of which are represented in Kent County. They appear as the facings of different plains or terraces, possessing very definite physiographic relations, as already described.

On purely lithologic grounds it is impossible to separate the three formations composing the Columbia group. The materials of all have been derived mainly from the older formations which occur in the immediate vicinity, mixed with more or less foreign material brought in by streams from the Piedmont Plateau or from the Appalachian region beyond. The deposits of each of these formations are extremely variable and change in general character according to the underlying formations. Thus materials belonging to the same formation may in different regions differ far more lithologically than the materials of two different formations lying in proximity to each other and to the common source of most of their material. Cartographic distinctions based on lithologic differences could not fail to result in hopeless confusion. At some places the older Pleistocene deposits are more indurated and their pebbles more decomposed than are those of younger formations, but these differences cannot be used as criteria for separating the formations, since loose and indurated, fresh and decomposed materials occur in each.

The fossils found in the Pleistocene deposits are far too meager to be of much service in separating them into distinct formations, even though essential differences between deposits may exist. It is the exceptional and not the normal development of the formations that has rendered the preservation of fossils possible. These consist principally of fossil plants that were preserved in bogs, although in a few places about Chesapeake Bay local Pleistocene deposits contain great numbers of marine and estuarine mollusks.

The Columbia group, as may be readily seen, is not a physiographic unit. The formations occupy wave-built terraces or plains separated by wave-cut escarpments, their mode of occurrence indicating different periods of deposition. At the bases of many of the escarpments the underlying Cretaceous and Tertiary formations are exposed. The highest terrace is occupied by the oldest deposits, the Sunderland, while the lowest terrace is made up of the youngest, or Talbot materials.

At almost every place where good sections of Pleistocene materials are exposed the deposits from base to top seem to be a unit. At other places, however, certain layers or beds are sharply separated from overlying beds by irregular lines of unconformity. Some of these breaks disappear within short distances, showing clearly that they are only local phenomena in the same formation, produced by contemporaneous erosion due to shifting shallow-water currents. Whether all these breaks would thus disappear if sufficient exposures occurred to permit the determination of their true nature is not known. An additional fact which indicates the contemporaneous erosive origin of these unconformities is that in nearby regions they seem to have no relation to one another. Since the Pleistocene formations lie in a nearly horizontal plane it would be possible to connect these separation lines if they were subaerial erosional unconformities. In the absence of any definite evidence that these lines are stratigraphic breaks separating two formations, they have been disregarded. Yet it is not improbable that in some places the waves of the advancing sea in Sunderland, Wicomico, and Talbot times did not entirely remove the beds of the preceding period of deposition over the area covered by the sea in its next transgression. Especially would deposits laid down in depressions be likely to persist as isolated remnants which later were covered by the next mantle of Pleistocene materials. If this is the case each formation from the Brandywine to the Wicomico is probably represented by fragmentary deposits beneath the later Pleistocene formations. In regions where pre-Quaternary materials are not exposed in the bases of the escarpments each Pleistocene formation near its inner margin probably rests upon the attenuated edge of the next older formation. Since lithologic differences furnish insufficient criteria for separating these late deposits, and since sections are not numerous enough to furnish distinctions between local interformational unconformities and widespread unconformities resulting from an erosion interval, the whole mantle of Pleistocene materials occurring at any one point is referred to the same formation. The Wicomico is described as including all the gravels, sands, and clays overlying the pre-Brandywine deposits and extending from the base of the Sunderland-Wicomico escarpment to the base of the Wicomico-Talbot escarpment. Perhaps, however, materials of Brandywine and of Sunderland age may underlie the Wicomico in places. In like manner the Talbot may occasionally rest upon deposits of the Brandywine, Sunderland, and Wicomico.

THE WICOMICO FORMATION.

The Wicomico formation receives its name from the Wicomico River on the southern Eastern Shore of Maryland where deposits of this age are characteristically developed.

Areal Distribution.

The oldest Pleistocene deposits of Kent County belong to the Wicomico, which, on account of its extensive development, is the most important formation in the region. It occupies the broad divide between Chesapeake Bay and Delaware River and is the surface formation over nearly two-thirds of the county. It conceals from view many of the Cretaceous and Tertiary formations which otherwise would be exposed over the divides and which now appear only along the valleys of the streams, where the Wicomico materials have been removed by erosion. The formation is extensively developed throughout the northern and central portions of the Atlantic Coastal Plain and probably extends into the Gulf region. It forms the surface material of a gently sloping plain ranging in elevation from 40 to 100 feet above sea level.

Character of Materials.

The materials composing the Wicomico formation are extremely varied both in size and mineral character. Boulders, gravel, sand, and loam are all present. These are usually well stratified, yet the lithologic characters of the strata change so abruptly that it is not possible to follow one bed for any great distance. Again, there is no definite sequence of the materials, although in general the coarser constituents are found near the base of the section while the finer form the capping. Fine sands may alternate with coarse boulder beds several times within a single section. Cross-bedding is also quite common.

The section exposed in the Betterton cliffs, given on a previous page, well illustrates the general character of the formation in its greatest development in this county. In the central portion of the divide between the Sassafras and Chester rivers, the Wicomico is from 15 to 20 feet thick and almost invariably has a well developed loam cap at the top and a gravel band, a few feet in thickness, at the base. This loam, in many respects, resembles the loess of the Mississippi Valley and constitutes the heavier soils of the upland. The gravel band at the base is exposed along the roads in scores of places where small streams are crossed.

The Wicomico materials found in almost every exposure have been largely derived from older deposits in the immediate vicinity, and the lithologic character of the formation changes from place to place according to the character of the contiguous older formations. In the northwestern portion of the county where a great deal of the Wicomico material has been derived from Upper Cretaceous and Eocene beds, the formation comprises considerable greensand. In the southeastern portion greensand is entirely absent, but the formation there contains light-colored sands derived from the Calvert. At places where there is little foreign material mixed with the locally derived débris it becomes somewhat difficult to draw the line between the Wicomico and the underlying Cretaceous and Tertiary formations. Usually, however, a stratigraphic break may be noted if there is a good exposure; if not, the harsher and more loamy character of the overlying materials indicates that they have been reworked and redeposited.

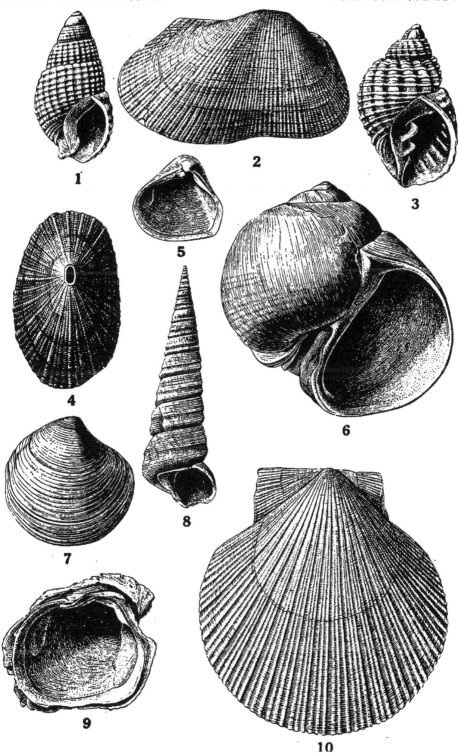

While the Wicomico was being formed as an offshore deposit streams from the adjoining land to the northwest were bearing in quantities of boulders, pebbles, sands, and loam. These were dropped when the streams entered the ocean, the larger particles first and the finer later. This sorting or arrangement is well shown in Kent County in that the size of the land-derived materials rapidly decreases from the northwest to the southeast. Large boulders and coarse pebbles are very common all over Sassafras Neck, but they gradually decrease both in size and number toward the southeast. Some of the boulders that occur in the Wicomico deposits in the northern portion of the county are very large. On Sassafras Neck many with a diameter of 2 feet or more are to be found. Nearly all the pebbles and boulders are composed of quartz or quartzite, but some of them are more complex mineralogically. In a ravine about one-half mile northeast of Galena some pebbles and boulders composed of peridotite, gabbro, gneiss, and quartz-mica schist were found. Similar boulders also occur elsewhere.

In the Potomac Valley near Washington boulders carrying glacial striae have been found in the Wicomico formation, but in Kent County no striated rocks have been observed. The great size of the boulders found here, however, and their occurrence with much finer materials furnish evidence of their transportation by floating ice.

The amount of loam present in the Wicomico is exceedingly variable. Wherever the loam cap is well developed the roads are firm and the land is suitable for producing grass and grain, but in regions where loam is present in small quantities, or absent altogether, the roads are apt to be sandy. The Wicomico on Sassafras Neck and on the divide between Sassafras and Chester rivers is characterized by its well-developed loam cap. In marked contrast with those regions is that portion of Delaware lying a short distance southeast of Kent County where there is little loam present and the surface is very sandy.

Physiographic Expression.

In Kent County the Wicomico possesses the features of a broad, flat plain forming the stream divides. On the west side of Chesapeake Bay it occurs mainly as narrow terraces occupying the lower portions of the stream divides and extending up the sides of the wider valleys. It is at many places separated from the Sunderland above and the Talbot below by well-defined escarpments. In Kent County the Wicomico–Talbot escarpment forms one of the most prominent topographic features of the region. It is best developed in the western portion of the county where it forms an almost continuous cliff 15 to 30 feet in height extending from Clinton Creek to Chestertown, passing near Hanesville, Melitota, Fairlee, Sandy Bottom, and Langford. It is a sharp rise that distinctly separates the low-lying Talbot plain from the upland Wicomico level. The same escarpment is continued from Chestertown to Millington usually at a distance of from one-half to one mile from the river. Along the Sassafras River it is less well developed though there it appears along the lower headlands between the tributary streams. These escarpments represent wave-cut cliffs formed during a period of submergence when the waters of Chesapeake Bay and its estuaries encroached upon the land to a greater extent than at present.

Paleontologic Character.

The Wicomico formation in Kent County has thus far furnished no fossils. In other regions plant remains and impure peat have been found in it. The plant remains have marked modern characteristics.

Strike, Dip, and Thickness.

The Wicomico formation is not at all places uniformly thick, owing to the uneven surface on which it was deposited. Its thickness ranges from a few feet to 50 feet or more. The formation dips down into the valleys and rises on the divides, so that its thickness is not so great as might be supposed from the fact that the base is frequently as low as 40 feet while the surface rises in places as high as 100 feet above sea level. Notwithstanding these irregularities it occupies as a whole an approximately horizontal position, with a slight southeasterly dip.

Stratigraphic Relations.

The Wicomico unconformably overlies all of the Cretaceous and Tertiary formations of the county. In places it may possibly overlie some of the Sunderland deposits which may be present beneath the Wicomico but, as has already been stated, the evidence for this is not conclusive. It is also in contact with the Talbot formation at many places along the Chesapeake Bay and its estuaries where the two formations are separated by an escarpment 10 to 20 feet in height, the base of which is from 38 to 45 feet above sea level.

THE TALBOT FORMATION.

The Talbot formation receives its name from the extensive development of deposits of this age in Talbot County.

Areal Distribution.

The latest formation represented in this region is the Talbot. It consists of gravels, sands, and loam in the form of a terrace that extends from tide to an elevation of 38 to 45 feet above sea level, where it is separated by an escarpment from the deposits of the Wicomico formation. It is best developed along Chesapeake Bay and the lower courses of the Chester River where it forms the surface material over a strip of land that has a width of between 3 and 7 miles. It also covers the lower portions of the minor stream divides along both the Sassafras and Chester rivers. The formation has an extensive development throughout the northern and middle portions of the Atlantic Coastal Plain.

Character of the Materials.

The materials composing the Talbot deposits are very similar in lithologic character to those found in the Wicomico formation. There is usually more loam present as compared with the gravel and sand than is found in the Wicomico, but the proportions of these constituents are extremely variable.

The following section, which is exposed near the town of Sassafras along a tributary of Sassafras River, illustrates the varied character of the material composing the Talbot formation:

SECTION NEAR SASSAFRAS.

	Ft.	In.
Coarse brown sandy loam..............	1	6
Fine gravel in matrix of coarse brown sand; no distinct bedding...........	4	6
Coarse brownish-yellow to buff sand with considerable fine gravel	2	10
Gravel band; pebbles of various sizes, not well sorted	0	5
Medium-fine brownish sand showing distinct cross-bedding, containing large broken masses of iron crust..........	8	0
Gravel band; pebbles small	3	6
Layer of iron crusts	0	10
Coarse sand, cross-bedded, containing a few drab-colored clay lenses..............	4	0
Indurated iron band	0	6
Coarse sand (thickness exposed)........	2	0
	30	1

Near Wilson Point wharf a section of the Talbot is exposed in which the material consists of weathered glauconitic sand. At Chestertown the Talbot contains beds of clay loam which have been utilized in the manufacture of brick.

Physiographic Expression.

The Talbot preserves the appearance of a terrace in the Kent County area better than the other Pleistocene formation. From the base of the Talbot-Wicomico escarpment the terrace slopes to the water's edge with few irregularities except where the surface is cut by streams.

Paleontologic Character.

The Talbot is the only one of the Pleistocene formations that has furnished any fossils in this region. On the east and west shores of Eastern Neck and on the bay shore of Eastern Neck Island, layers of clay containing plant remains have been found. The vegetable débris has been rather finely comminuted though no doubt careful collecting would reveal the presence of determinable species, such as have been studied from other places on Chesapeake Bay. Elsewhere large fossil cypress stumps, in an upright position, have been exposed by the cutting action of the waves, but no such fossils have been observed in this county. At Cornfield Harbor, near the mouth of Potomac River, the formation has yielded a great number of molluscan shells, representing a varied fauna of marine and brackish-water origin.

Strike, Dip, and Thickness.

The thickness of the Talbot formation is extremely variable ranging from a few to 40 or more feet. The unevenness of the surface on which it was deposited has in part caused this variability. The proximity of certain regions to mouths of streams during the Talbot submergence also accounts for the increased thickness of the formation in these areas.

Stratigraphic Relations.

The Talbot rests unconformably, in different portions of the region, upon various formations of Cretaceous and Tertiary age. It may in places rest upon deposits of Sunderland or Wicomico age, although no positive evidence has yet been found to indicate such relations to the older Pleistocene formations. The deposits occupy a nearly horizontal position, with perhaps slight slopes toward Delaware and Chesapeake bays on the two sides of the county, but the amount of slope is too small to be accurately determined. The Talbot was at some places deposited upon a very irregular surface. Great irregularities, now concealed, no doubt exist elsewhere in the surface upon which the Talbot materials were deposited.

THE RECENT DEPOSITS.

In addition to the two Pleistocene terraces already discussed, a fifth is now being formed by the waters of the rivers and the waves of the estuaries. This terrace is present along the water's edge, extending from a few feet above to a few feet below tide. It is the youngest and topographically the lowest of the series. Normally it lies at the base of the Talbot terrace from which it is separated by a low scarp. In the absence of the Talbot the Recent terrace may be found at the base of the Wicomico terrace in which case the separating scarp will be higher. The Recent materials consist of peat, clay, sand, and gravel and these are deposited in deltas, flood-plains, beaches, bogs, dunes, bars, spits, and wave-built terraces.

INTERPRETATION OF THE GEOLOGICAL RECORD

The area in which Kent County lies has undergone many changes throughout past geologic time, some of which can be readily interpreted by the character of the deposits and their physical relations. The region has alternately been submerged and elevated, and deposition of materials has frequently been succeeded by erosion. At certain times the entire county was beneath the water and received deposits; at other times it was land and was degraded by surface streams; at still other times the shore line crossed the county so that part of it was in the zone of denudation and part of it in the zone of deposition. The erosion intervals are indicated by erosional unconformities, while the beds of various materials represent periods of submergence. Further, the physical conditions prevailing during the ages of sedimentation are revealed by the lithologic character of the beds and their included organic remains.

The floor upon which the Coastal Plain deposits were laid down is a great mass of crystalline rocks of pre-Cambrian and early Paleozoic age. These crystallines do not appear at the surface in this region, nor have they been reached by any deep-well borings.

SEDIMENTARY RECORD OF THE LOWER CRETACEOUS.

The earliest of the known unconsolidated deposits lying upon the floor of crystalline rocks belong to the Patuxent formation of the Potomac group. It does not appear at the surface within Kent County, but has been reached by the deep-well boring at Middletown, Delaware. It outcrops a few miles to the northwest, in Cecil County, and probably underlies this entire county. It indicates a submergence of the Coastal Plain of this region of sufficient extent to cover the whole area with shallow water. The cross-bedded sands and gravels furnish evidence of shifting currents, as do also the rapid changes in the character of the materials, both horizontally and vertically. The presence of numerous land plants in the laminated clays shows the proximity of the land.

The deposition of the Patuxent formation was ended by an uplift which brought the region above the water and inaugurated an erosion period which persisted long enough to permit the removal of a vast amount of material. To the south a submergence, during which the Arundel formation was laid down, and a re-elevation occurred before the area of this county was again depressed beneath the water level. Physical conditions similar to those which had prevailed during Patuxent time existed during this period of submergence, in which the Patapsco formation was laid down.

After the deposition of the Patapsco formation the region again became land through an upward movement which drained all of the previously existing estuaries and marshes. Erosion at once became active and the Patapsco surface was dissected.

SEDIMENTARY RECORD OF THE UPPER CRETACEOUS.

A downward land movement again submerged the greater portion of the region, leaving only a very narrow strip of Patapsco deposits above water. The Raritan formation was now deposited under conditions very similar to those which had existed during the previous submergence. Raritan deposition was terminated by an uplift which again converted the entire region into land. A long period elapsed before a re-submergence, so that the streams were able to extensively erode the recently formed deposits.

The extensive development of shallow-water deposits, everywhere cross-bedded and extremely variable in lithologic character, and the presence throughout of land plants furnish some evidence that Raritan sedimentation took place, not in open ocean waters but in brackish or fresh-water estuaries and marshes that were indirectly connected with the ocean, which may have at times locally broken into the sea. Some land barrier east of the present shore line probably existed and produced these conditions, but its position and extent cannot be determined.

The period during which the Magothy deposits were formed was a period of transition from the estuarine or fresh-water conditions of the Patapsco and Raritan periods to the marine conditions of the Matawan and Monmouth periods. The lithologic characters of the materials as shown by their great variability, the coarseness of the sands and gravels, and the cross-bedding all suggest conditions similar to those of the former periods. On the other hand, the occasional pockets of glauconitic sand and the presence of marine invertebrates suggest the marine conditions of the later Cretaceous periods. The probability is that over most of the area where Magothy deposits are now found Potomac conditions prevailed during the greater part of the period and in some places perhaps during the whole period, but that occasionally, through the breaking down of the land barriers which had kept out the ocean, there were incur-

FIG. 1.—VIEW SHOWING HILLSIDE EROSION AT UPPER EDGE OF THE WICOMICO-TALBOT
SCARP IN KENT COUNTY.

FIG. 2.—VIEW SHOWING HILLSIDES WITH HARDWOOD FOREST, BORDERING MARSH LAND
NEAR STILL POND.

sions of sea water, bringing in marine forms of life. Thus far there is no evidence that they occurred anywhere except in New Jersey.

At the close of the Magothy period the region was uplifted and a period of erosion was inaugurated. During this erosion interval comparatively small amounts of material were removed. In some places it is impossible to establish definitely any erosion break between the Magothy and the Matawan. This may be because the erosion interval was comparatively short or because the elevation of the land above the water was so slight that it did not permit the streams to cut channels in the recently formed deposits.

Not until late Cretaceous time did a downward movement occur of sufficient extent to permit the ocean waters to transgress widely over this region. During the Matawan and Monmouth epochs all of Kent County, except perhaps a small portion in its northwestern corner, was depressed beneath the ocean waters. The streams from the low-lying land evidently carried into the ocean at this time only small amounts of fine sand and mud, which afforded conditions favorable to the production of glauconite and permitted the accumulation of the greensand beds that are so characteristic of the Upper Cretaceous epoch along the Atlantic border. During this time very slight changes took place along the continental border, although elevation was probably proceeding slowly, as the Monmouth formation is found outcropping farther and farther southeastward.

After the deposition of the Rancocas formation in adjoining territory in Delaware and New Jersey upward land movements again caused the shore line to retreat eastward, but to what point is not definitely known. In areas lying farther north in New Jersey, deposition still continued in some places, for the Rancocas is there overlain by another later deposit of Cretaceous age. If such deposits were ever formed within the limits of Kent County they have either been removed or are concealed from view by later formations which have overlapped them.

SEDIMENTARY RECORD OF THE EOCENE.

During early Eocene time a portion of this area again became a region of deposition through a subsidence which carried it beneath the ocean waters. This Eocene ocean seems to have transgressed the Rancocas surface, as Eocene deposition took place immediately upon the Monmouth formation in many places along Sassafras River and its tributaries. The Eocene waters probably did not cover that portion of Delaware adjoining Kent County for near Middletown and Townsend, Delaware, the later Calvert deposits are in contact with the Rancocas.

The conditions that prevailed during the time of the deposition of the Aquia formation must have been very similar to those existing during late Cretaceous time. The presence of great quantities of glauconitic material indicates quiet water where foraminifera abounded and where only fine terrigenous detritus was being carried in small amounts by streams from the land. The waters were also well suited for marine life of higher types, and numerous pelecypod and gasteropod fossils occur in the deposits.

SEDIMENTARY RECORD OF THE MIOCENE.

The Eocene deposits are unconformably overlain by the Calvert formation. The unconformity indicates that an erosion interval succeeded Eocene deposition, during which the area was above water and the streams of the region were cutting drainage channels in the Eocene deposits. A subsequent depression of the district submerged all that portion of the county lying southeast of a line drawn from Sassafras to Kennedyville. At this time all of the land to the west must have been worn down to such an extent that the streams which drained it had very little force. Fine sands and mud were carried into the ocean and laid down as an offshore deposit, but no coarse materials were brought in. Diatoms lived in abundance in the waters near the shore and as they died their siliceous tests dropped to the bottom. Although diatoms are extremely small, yet their remains form a very considerable portion of the Calvert deposits, and in places beds several feet in thickness are found composed almost entirely of their tests. The Calvert deposits must therefore represent a very long period of time. The waters also abounded in other forms of life, particularly corals, pelecypods, gasteropods, and fishes, although all the main groups of marine animals are represented.

After the deposition of the Calvert formation most if not all of this region remained above water for a long period, during which those portions of the Atlantic Coastal Plain that lie farther east and south were alternately submerged and uplifted. Two Miocene formations, not represented in this area, are developed in those districts. During this time erosion was active, and much material was removed by the streams that meandered across the region.

Sedimentary Record of the Brandywine Formation.

The erosion interval that followed Calvert deposition was finally terminated by a more extensive submergence, which carried the whole region beneath the waters of the ocean and at the same time elevated the adjoining land through a southeastward tilting of the continental border. This tilting rejuvenated the rivers and they were enabled to carry much coarser materials than they had borne during Eocene and Miocene time. As a result the entire submerged region near the shore was covered with a mantle of coarse gravel and sands, while the finer materials were carried out to sea beyond the confines of Kent County. These deposits constitute the Brandywine formation. The thickness of this formation, in view of the coarseness of the materials, indicates that this submergence was not of long duration. This material was deposited on a gently sloping surface, probably similar to the present continental shelf. In time upward-moving forces became dominant and the entire Coastal Plain was again raised above the water. When the region was uplifted the recently deposited material formed a broad, nearly level plain, which extended from the Piedmont Plateau in a gradual slope to the ocean. Erosion succeeded deposition and large quantities of the Brandywine material were removed. During this erosion interval streams rapidly cut into the Brandywine and earlier formations. Over Kent County the Brandywine plain was entirely destroyed, while in other places the tributary streams succeeded in isolating large portions, which remained as outliers.

SEDIMENTARY RECORD OF THE PLEISTOCENE.

During the next depression, which occurred in Pleistocene time, the Sunderland deposits were formed. The depression was not great enough to carry all portions of the Coastal Plain beneath the water, and only those regions which now have an elevation less than 180 feet above sea level were submerged. All of Kent County seems to have been submerged. The materials that were carried in by the streams and deposited in the ocean, there to be re-sorted by the waves, indicate that the relation of the land to the sea must have been about the same as during Brandywine time. In the valleys which had been carved out by the streams during the erosion interval following the Brandywine period the deposits formed were much thicker than on the former stream divides. Had the period of submergence been a long one the old stream valleys must have been obliterated. That the Sunderland period, like the preceding, was comparatively short may be inferred from the thin layer of sediments which accumulated over the submerged region.

An elevation sufficient to bring the entire area above water permitted the streams to extend their courses across the newly-formed land and in a short time the Sunderland deposits were extensively eroded. A portion of those that remained after this period of denudation were destroyed by the waves, when a gradual subsidence again permitted the ocean waters to encroach upon the land. In this submergence the regions now lying above 100 feet were not covered with water; hence a considerable part of the Coastal Plain remained as land. At this time the Wicomico sea cut cliffs along the shore and these now appear as escarpments whose bases are at an elevation of 90 to 100 feet above sea level. Streams of considerable velocity and volume brought down gravel and sand, which the waves spread over the ocean bottom. The coarser materials were dropped near the shore, while the finer were carried farther out to sea. This accounts for the fact that the gravel of the Wicomico formation is larger and more abundant in the northwestern portion of the county than in the southeastern portion.

During the time that the Wicomico formation was being laid down the country to the north was covered by the glacial ice sheet. A great deal of ice evidently formed along the streams that were bringing in the Wicomico materials, and at times large masses were broken loose and floated down to the ocean. These ice masses carried within them boulders, frequently of large size, which were dropped as the ice melted, and in this way the boulders that are found in Wicomico deposits, mixed with much finer deposits, reached their present positions. Some of these ice-borne boulders included in the Wicomico deposits found elsewhere show their glacial origin by numerous striae. Toward the close of Wicomico time an upward land motion caused the ocean to retreat gradually again and at the same time checked the velocity of the streams through a landward tilting, by which the lower courses were elevated to a greater degree than the upper courses. The streams with less carrying power were then unable to transport coarse materials and as a result the upper beds of this formation are composed principally of fine sand and loam.

During the succeeding erosion interval the principal streams that are now present in this region developed, in large part, their main and lateral channels as they now exist.

The lower courses of Sassafras and Chester rivers in their present form date from this time. Before the next subsidence all of these streams had cut through the Wicomico deposits and opened wide valleys in the old channels. With later submergence the water entered these valleys, converting them into wide estuaries or bays. The greater portion of the region was not submerged; those areas that now have an elevation more than 40 feet above sea level remained as land. In the estuaries and bays the Talbot deposits were laid down. In Kent County the Chesapeake Bay shore line at this time extended irregularly from Churn Creek to Chestertown, and along this shore the waves were sufficiently strong to cut sea cliffs at many exposed points. These remain as escarpments and may be plainly seen at several points, particularly near Langford and Melitota. The waters of Chesapeake Bay advanced up the valleys of the various streams, forming broad estuaries in which sedimentation took place. Although the Bay was then, as now, merely an arm of the ocean, yet the waves were of sufficient magnitude to cut sea cliffs at many places. In this region some of these old sea cliffs can be traced continuously for several miles as escarpments, in places 15 to 20 feet high. During this period of submergence the waters of Chesapeake Bay extended far enough inland to permit deposition in areas as far east as Sassafras on Sassafras River and Millington on Chester River.

The Talbot materials closely resemble those of the Wicomico formation, which indicates similar conditions during the two periods. Along the shore at some places marshes were formed in which an accumulation of vegetable debris took place, as in the swamps on Eastern Neck, which were produced at this time.

The Talbot stage of deposition was brought to a close by an uplift, as a result of which the shore line once more retreated and the previously submerged regions were drained. When this elevation occurred the region that emerged from the sea appeared as a broad terrace about the borders of the Wicomico plain, above described. During this time of uplift the streams again became active and rapidly removed large quantities of the loose material that had just been deposited. The land after the uplift undoubtedly stood at a higher elevation than at present, so that the material recently deposited formed a larger addition to the continent than would appear from the present outlines of the Talbot formation. Although a comparatively short period has elapsed since the Talbot deposits were converted into land, yet already in many places the streams have succeeded in cutting through these to the underlying beds.

The last event in the geologic history of the region was a·downward movement, which is still in progress. It is this which has produced the estuaries and tidewater marshes that form so conspicuous features of the present topography. The movement is very slow and in many places has not kept pace with the filling process which is very noticeable in certain regions of the Coastal Plain. Many of the estuaries are not now navigable as far inland as they were a century ago. Deposition is very active in the estuaries, as nearly all the material brought down by the streams from the land is dropped in their quiet waters. The following statements indicate the amount of change that has taken place within the Delaware River in recent times and probably similar changes have occurred in almost all of the tidewater estuaries of the region. From 1841 to 1881 Delaware River between Reedy Island and Liston Point increased its mean width 411 feet, 285 feet on the New Jersey side and 126 feet on the Delaware shore. During this same period certain portions of this area have been deepened while certain others have been shoaled. Except in the region of Liston Point the river bed shows an excess of shoaling over deepening. The region includes an area of 15 square miles and shows an excess of filling of 8,096,150 cubic yards, representing an average decrease in depth of 0.4 foot in forty years. (Rept. U. S. Coast and Geodetic Survey for 1884, Appendix 12, pp. 433-434.)

THE MINERAL RESOURCES OF KENT COUNTY

BY

BENJAMIN L. MILLER

INTRODUCTORY.

The mineral resources of this region are not extensive nor extremely valuable, yet Kent County contains some deposits of economic importance, although they have not been very largely worked.

THE NATURAL DEPOSITS

THE CLAYS.

The Pleistocene formations of this region contain a number of clay beds, some of which are available for the manufacture of brick and tile. In Chestertown the surface loam of the Talbot formation has been utilized for the making of brick. The material is used to a depth of about 4 feet. No doubt much of the surface Talbot loam on the broad low-lying flats bordering the lower Chester River and Chesapeake Bay would prove equally serviceable for the manufacture of ordinary brick and tile.

On the Uplands, the Wicomico loams cover extensive areas and would, in many places, be suitable for brick. In the vicinity of Philadelphia and Washington, and in many places in Virginia these Wicomico argillaceous loams have been extensively utilized for this purpose.

THE SANDS.

In the Pleistocene and Miocene formations there are numerous and extensive beds of fine quartz sands. The sand from these beds has been used locally for building purposes, but no large openings have been made in any of the deposits. In many places in the county an unlimited amount of sand of excellent quality for building purposes could be obtained.

THE GRAVELS.

The Wicomico and Talbot formations contain many beds of gravel that is suitable for road-making, and in a few places these beds have been worked. In many places the deposits contain enough ferruginous clay and sand to cause the gravel to pack well and to make a firm road bed. There are, in the region, many beds not yet opened which would yield a good quality of gravel for road making. In the upland portion of the county gravels are found almost universally beneath the loam cap of the Wicomico, and these have been used here and there for local purposes. At some places these gravel beds contain very little sand or clay and consequently are not well suited for roads; at others there is considerable iron oxide and sandy clay mixed with the gravel and it has considerable value as road metal.

THE MARLS.*

The Monmouth and Aquia formations contain considerable glauconitic and calcareous materials. It is well known that glauconitic marl has considerable value as a fertilizer. Similar deposits have been extensively worked in New Jersey, where the importance of utilizing the marls has long been recognized. The marls of Kent County seem to be somewhat inferior in quality to many of the New Jersey deposits, for analyses show a smaller percentage of the potassium compounds, yet the results obtained by the use of the Delaware and Maryland marls are said to have been very satis-

* It should be understood that the marls of the coastal plain differ widely in character and origin from the marls used in Michigan for the manufacture of cement. The marls of Maryland are not suitable for such purposes.

factory. In the early part of the last century many marl pits were opened in Newcastle County, Delaware, and Cecil County, Maryland, where these glauconitic beds either appear at the surface or under thin cover of later deposits. These marl pits were located near Silver Run and Drawyer and Appoquinimink creeks in Delaware, and along Bohemia Creek and Sassafras River and their tributaries in Maryland. Marl was obtained also at a few places near Sandy Bottom. Analyses made a long time ago by the Delaware Geological Survey show from 7 to 9 per cent of potassium. In places where the marl can be obtained at low cost such a percentage of potassium would seem to justify the opening of marl pits for local use.

THE BOG-IRON ORE.

In many places on the Eastern Shore of Maryland deposits of bog-iron ore are found in the swamps and marshes bordering the estuaries. Conditions have long been favorable for its accumulation and considerable deposits have formed in some of the lower counties where conditions are still favorable for its formation. In the early history of the region, many of these deposits were used as a source of iron though, at the present time, they could not be worked with profit. In Kent County the iron ores are of minor importance and, so far as is known, have never been utilized. Thin layers of ore are exposed, however, on Eastern Neck and in the vicinity of Langford Bay where the waves have removed the over-. lying materials and no doubt there are other places where the iron ore is developed though not exposed.

THE WATER RESOURCES

The water supply of Kent County available for use is found in the streams and wells of the district. As the county contains no large cities the streams are not used as sources of public water supply. They are, however, used to furnish water power in some places, as has been already mentioned.

SURFACE WATERS.

The two large streams, Sassafras River and Chester River, which form respectively the northern and southern boundaries of the county, are tidal estuaries and their waters are consequently brackish and more or less charged with organic matter and are therefore unsuited as sources of potable water. The small streams are all short and expand almost immediately into estuaries or marshes. The amount of flow is limited, the water is charged with organic matter, and practically all receive more or less drainage from inhabited areas and are therefore extremely liable to pollution. None are, nor should be, utilized as sources of domestic or municipal supplies.

UNDERGROUND WATERS.

ARTESIAN WATERS.

The absence of large centers of population or industrial enterprises requiring large quantities of water has limited the number of drilled wells in Kent County, since shallow dug, or driven well furnish ample supplies for domestic or farm purposes. At Chestertown two of the wells at the ice plant, between 160 and 170 feet deep, are thought to draw water from a bed in the Monmouth. The water only rises to within 30 feet of the surface but the supply is large. A well about 1½ miles southwest of Morgnec is 224 feet deep and is drawing from the Matawan. The water is soft and rises to within 10 feet of the surface.

Two wells at Rockhall found water at about 345 feet that is thought to come from the Magothy, being correlated with the lower Magothy level in the Chestertown well. These wells have a fair flow, the exact amount unknown, but the water is so highly charged with iron that it is not palatable.

At the Chestertown Water Works two attempts have been made to secure water from deep wells, but although water was found in both wells the results were not satisfactory and so the wells were abandoned. The log of the second and deeper well is given below.

DEEP WELL OF CHESTERTOWN WATER WORKS.

(Put down winter of 1908-1909. J. H. K. Shannahan Company, Contractors.)

Pleistocene.	Feet
Soft yellow clay	0–6
Soft yellow marl containing shells	6–60

Eocene.
Aquia formation.

Soft gray marl containing shells	60–113
Soft black marl, hard boulders	113–129

Upper Cretaceous.
Monmouth formation.

Hard and soft marl alternating from green to black..	129–150
Hard dark brown sand	150–200

Matawan formation.

Gray and black sand, water bearing, pumped 15 gals.	200–230
Gray clay and sand	230–251
Soft gray sand rock	251–257
Hard black sandy clay	257–268

Magothy formation.

Soft black loamy micaceous clay	268–332
Soft coarse white sand, water bearing, tested about 20 gallons per minute	332–335
Soft lead-colored clay	335–340
Soft coarse white sand, water bearing, no test	340–344

Raritan formation.

Soft clay alternating red and white	344–355
Soft sandy clay alternating red and white	355–390
Reddish sands, grains loose and free, water bearing..	390–395
Soft sandy red clay	395–421
Hard red clay	421–480
Soft rock (sandstone?)	480–480½
Sand, traces of water	480½–481½
Soft gray clay	481½–492
Very hard rock	492–492½
Tough sticky gray clay	492½–540
Hard gray sandy clay	540–550
Free white sand, water bearing, tested 80 gallons per minute	550–581

Lower Cretaceous.
Patapsco formation.

Soft gray sandy clay	581–625
Very hard red and white sandy clay	625–632

Hard boulder 632–632½
Tough light pink clay...........................632½–648
Tough red clay 648–700
Gray sandy clay, alternated hard and soft.......... 700–706
Coarse white sand, trace of water................ 706–713
Tough purple clay 713–750
Tough red clay 750–955
Patuxent formation.
Soft purple clay containing hard boulders.......... 955–981
Very hard purple clay.......................... 981–1002
Coarse reddish sand, quite free. Flows 14 gallons per
    minute at +2 feet......................... 1002–1004
Soft gray clay................................ 1004–1023
Red clay, somewhat hard....................... 1023–1050
Soft gray clay................................ 1050–1056
Very hard gray clay........................... 1056–1060
Soft red clay 1060–1100
Soft gray sandy clay containing boulders........... 1100–1108
Soft gray sandy clay.......................... 1108–1110
Large boulders 1110
Soft gray sandy clay.......................... 1110–1135
Coarse sand, water bearing. Flows 50 gallons per
    minute. Very salty........................ 1135

The first well stopped at 583 feet near the base of the Raritan, where a supply was encountered which overflowed 20 gallons per minute. The second well passed through to the Patuxent sands at a depth of 1135 feet and encountered a flow of about 50 gallons per minute with a head 2 feet above the surface, but this water was too salty to use. The abandonment of these two wells, or more strictly the shallower; since the water in the deeper well was unfit to use, demonstrates the need of a change in well-drilling methods. By the system in use in the Coastal Plain of Maryland the driller merely washes out a hole in the ground, hammering down castiron pipe until he strikes a stratum which yields a flow or a good head of water. The small water zones that may be penetrated escape his notice, and since he knows no way to add them to the large flow which he hopes to strike he seldom keeps accurate records of the beds passed through or the water they contain.

Soft yellow clay ....

Soft yellow marl containing shells ....

Soft gray marl containing shells............................

Soft black marl, hard boulders.............................

Hard and soft marl alternating from green to black............

Hard dark brown sand ....

Gray and black sand, water bearing, pumped 15 gals.............
Gray clay and sand ....
Soft gray sand rock ....
Hard black sandy clay ....

Soft black loamy micaceous clay ....
Soft coarse white sand, water bearing, tested about 20 gals. per min
Soft lead colored clay ....
Soft coarse white sand, water bearing, no test...............
Soft clay alternating red and white.........................
Soft sandy clay alternating red and white....................
Reddish sands, grains loose and free, water bearing............

Soft sandy red clay......................................

Hard red clay ....
Soft rock (sandstone?) ....

Sand, traces of water ....

Soft gray clay ....
Very hard rock ....

Tough sticky gray clay...................................

Hard gray sandy clay ....
Free white sand, water bearing, tested 80 gals. per min...........

Soft gray sandy clay ....

Very hard red and white sandy clay.........................
Hard boulder ....
Tough light pink clay ....

Tough red clay ....

Gray sandy clay, alternated hard and soft.....................
Coarse white sand, trace of water..........................
Tough purple clay ....
Tough red clay ....
Soft purple clay containing hard boulders....................
Very hard purple clay ....
Coarse reddish sand, quite free. Flows 14 gals. per min. at + 2 f
Soft gray clay ....
Red clay, somewhat hard ....
Soft gray clay ....
Very hard gray clay ....
Soft red clay ....
Soft gray sandy clay containing boulders ....
Soft gray sandy clay ....
Large boulders ....
Soft gray sandy clay ....
Coarse sand, water bearing. Flows 50 gals. per min. Very salty...

Fig. 1.—Section of the Deep Well at Chestertown.

Referring to the section of the Chestertown well, it will be seen that there was a water level in the Eocene (the one from which the shallowest ice-plant well at that place draws), one in the Monmouth (the level which supplies the two other wells at the ice plant), a lower horizon in the Matawan, two in the Magothy, and three in the Raritan. But the finished well draws only from the last and yielded an insufficient supply.

The idea of drawing from several levels at once has perhaps occurred rather vaguely to those who have had to suffer most from this method of drilling, but the only way apparent to do this is by leaving the portions of the well opposite the water-bearing strata uncased. This is impossible because of the objectionable sand and silt that would be pulled up by the pumps, although in some deep wells in more consolidated material it is possible to leave the well uncased after the loose, near-surface materials have been passed.

The drillers in California were confronted with the same problems, sharpened considerably, in that strata were looser, filled with larger boulders, and were much thicker. Then, too, since the wells were sunk with the view of obtaining water for irrigation an enormous supply was demanded. These impelling reasons led to the development of a special type of well construction known as the "stovepipe" well which is discussed by C. S. Slichter.* Short lengths of large casing are forced down by hydraulic jacks, accurate records of water zones are kept, and after a sufficient number are penetrated to yield the desired supply the casings are slit or perforated at the desired levels by appropriate tools of which a considerable variety are in use in the West. To give an idea of the amount of water yielded by wells of this construction several yields of California wells are noted. From wells averaging 250 feet in depth 300,000 to 2,000,000 gallons a day have been pumped, while several deeper wells in southern California 500 to 700 feet deep flow 3,000,000 gallons daily. It may not be possible to duplicate these

* U. S. Geol. Survey, Water Supply Paper No. 110, pp. 32-36, 1905.

FIG. 1.—VIEW SHOWING TALBOT-WICOMICO SCARP, TALBOT SURFACE IN FOREGROUND, ONE MILE EAST OF SANDY BOTTOM.

FIG. 2.—VIEW AT THE SAME LOCALITY AS ABOVE BUT FROM THE WICOMICO SURFACE LOOKING DOWN ON THE TALBOT PLAIN. THE TOP OF THE SCARP MAY BE SEEN RUNNING ACROSS THE MIDDLE OF THE ILLUSTRATION.

yields in the East, but some such method would unquestionably have saved the Chestertown well and would greatly increase flows in other wells in the Coastal Plain at present dependent upon one-water bed.

The adoption of this method would necessitate a few radical changes in rigging. The hydraulic jacks might not be absolutely necessary, but the advantages of their use are so numerous that their inclusion in the rig would greatly increase its efficiency. Perhaps the greatest difficulty, however, aside from that of overcoming prejudice and custom, would be that of securing the proper casing in the East, although the sections are very simply constructed and could easily be copied.

Summarizing the artesian prospects in Kent County, it may be said that in the western part of the county the results of deep drilling have not been entirely satisfactory. In the vicinity of Millington fair supplies of hard water can be obtained at less than 125 feet, but the head is too low to give flows. However, the water rises to within less than 10 feet of the surface and can be easily pumped. Toward the shore of Chesapeake Bay it may be necessary to drill from 250 to 400 feet. Flows have been obtained on low ground near the Bay. Other artesian wells could doubtless be obtained by drilling to the same water horizons. The depth required can be estimated by adding 20 to 30 feet for each mile toward the southeast from known wells, or subtracting a like amount for each mile toward the northwest. The head is only a few feet above sea level, and flowing wells cannot be obtained except on the lower portion of the Talbot plain. On the higher ground the water should rise near enough to the surface to be pumped.

With a few exceptions the artesian wells of Kent County have obtained satisfactory water, but at Rockhall a 400-foot well encountered water high in iron, and at Chestertown the deep well procured salt water. These facts suggest that deep drilling may prove unprofitable, although elsewhere throughout the Coastal

Plain of Maryland the Lower Cretaceous water horizons have yielded large supplies and usually of good quality.

Springs.—Aside from small springs at various points and liable to more or less seasonable fluctuations there is one of good size along the Sassafras River at Betterton. This spring, known as the Idlewhile, has attracted considerable attention and is extensively advertised by the owner of the Idlewhile Hotel. The spring has a flow of about 25 gallons per minute and is reported to have had a constant volume during the last 40 years. It emerges in a small depression near the shore and the water probably comes from a sand bed in the Magothy formation. The construction of a wall about the spring and of a small house over it excludes dirt and surface water.

Shallow Wells.—The majority of the inhabitants of Kent County utilize shallow wells for their water supply since the water is usually obtainable in sufficient quantities for domestic or farm use at inconsiderable depths, is generally of good quality, and because of the equably distributed rainfall is dependable at all seasons of the year.

In the lower areas along the Bay and up the Chester River, as at Melitota, Tolchester, Sandy Bottom, Crosby, Edesville, Rockhall, and as far up the river as Millington, variable but usually sufficient amounts of water are found in the Talbot formation at depths ranging from 8 to 25 feet. Naturally these wells exhibit a variety of conditions reflecting their local environment, since the shallow water table is the direct result of downward seepage from the rainwater falling on the surface of the ground. In some places the water is pure and wholesome and free from organic or mineral matter. Elsewhere the water may be so high in iron or organic matter as to be unfit for use. The Talbot water is thus a very accessible supply and usually ample and of good quality, but very susceptible to local surface conditions and also liable to marked fluctuations during especially wet or dry seasons.

The broad level surface of the Wicomico terrace which forms the central and eastern part of the county comprises a thin mantle of sand and loam which like the Talbot stores the water that falls as rain on its surface. The water table is generally somewhat lower than on the Talbot terrace and the wells must be sunk somewhat deeper, striking their first water zone at the base of the Wicomico formation at depths varying with the surface topography and ranging from but 12 feet at Worton to the more common depth around 30 feet.

The Wicomico water, like the Talbot, is accessible and usually ample and of fair quality, although frequently hard. The older geological formations, already mentioned in the introductory paragraph on the geology of the county, lie so near the surface that they are readily tapped by comparatively shallow wells. In the northern part of the county along Sassafras River and in the northwestern part along the Bay the wells penetrate the Upper Cretaceous formation. At Betterton, where the wells vary in depth from 40 to 80 feet, an ample supply of good water is obtained from the Magothy formation. A well at Coleman, 70 feet deep, draws from this same horizon.

In the region underlain by the Aquia formation of the Eocene, it is only necessary to go to shallow depths to obtain Eocene water. At Galena three wells at different elevations strike Aquia water at from 40 to 60 feet. At Kennedyville, Morgnec, and Sandy Bottom this same water horizon is found at from 50 to 65 feet. At Chestertown the public supply wells penetrate this zone at from 58 to 70 feet, while at Millington, where the surficial formations are thick and are underlain by the Calvert it was necessary to go down 100 feet to strike the Aquia water zone. All the Eocene wells have a noticeable head, the Kennedyville well rising to within 4 feet of the surface. This Eocene water seems to be consistently hard but not otherwise objectionable. There should be no difficulty in finding this water in the southern part of the county and it should be especially valuable since it is not deeply buried and because it will be more dependable and not so easily depleted a supply as the surface waters of the Pleistocene, and by proper locating the water should be brought within easy pumping distance of the surface.

The public supply wells at Chestertown and one of the wells at the ice plant, 99 feet deep, probably all draw from the Eocene, although at different levels.

As previously mentioned, the southeastern part of Kent County is underlain by the Calvert formation of the Miocene which in Southern Maryland and the other lower Eastern Shore is a most important artesian horizon. It is unimportant in Kent County but is sometimes utilized by shallow wells in this part of the county.

Number	Location	Owner or Tenant	Altitude	Depth	Diameter	Length of casing	Depth to principal supply	Geologic horizon	Depth to subordinate supply	Head	Volume of flow	Yield by pumping	Character of water	Date drilled
1	Chestertown	Chestertown Water Co...	15	583	8	583	583	Raritan	...	+2	20	100	...	...
2	Chestertown	Chestertown Water Co...	15	1135	8	1135	1135	Patuxent	Cased off	+2	50	...	Salt	1901
3	Chestertown	R. G. Nicholson...	35	100	3	100	100	Aquia	Cased off	-30	...	...	Hard	...
4	Chestertown	R. G. Nicholson...	35	160	6	160	160	Monmouth	Cased off	-30	...	...	Hard	...
5	Chestertown	R. G. Nicholson...	35	170	6	170	170	Monmouth	Cased off	-30	...	...	Hard	...
6-19	Chestertown (14 wells)	Chestertown Water Co...	15	58-70	Five 3, Nine 2½	58-70	58-70	Aquia	...	-40	...	...	Hard	...
20	Chestertown, 3 mi. NE	H. Klinefelter	75	224	4½	......	223	Magothy	...	...	...	...	Soft	...
21	Galena	Catholic Church	65	61	3' 8"	Bricked	61	Aquia	...	-58	...	...	Hard	1860
21	Galena	Davis Bros.	...	60	1½	60	60	Aquia	...	-40	...	160	Hard	1916
22	Kennedyville	Penn. R. R.	...	65	6	65	65	Aquia	...	-15	...	11½	Soft	1880
23-25	Massey (3 wells)	Penn. R. R.	64	48-55	2	48-55	48-55	Calvert	...	-10	...	85	Soft	1909
26	Millington	J. P. Ahern	20	102	4	52	...	Aquia	...	-4	...	...	Hard	1907
27	Millington	Central Hotel	27	105	3	30	...	Aquia	...	-4	...	...	Hard	1907
28	Millington	J. E. Higgman	21	102	4	...	...	Aquia	...	-4	...	...	Hard	1907*
29	Millington	W. H. Soper	10	99	4	...	...	Aquia	...	-40	...	...	Hard	1907
30	Morgnec, 1½ mi. SW	....	45	224	4½	223	...	Matawan	...	-40	...	...	Soft	...
31	Rockhall, ½ mi. SE	Rockhall Canning Co...	3	300	2	300	300	Magothy	...	+10	2	...	Hard	...
32	Rockhall, ½ mi. SE	Rockhall Canning Co...	3	345	2	...	345	Magothy	...	+10	6	...	Hard	...
33	Rockhall, 1½ mi. S	G. E. Leary & Son...	3	400	?	400	400	Magothy	...	+8	5	...	Hard	...
34	Rockhall, 1½ mi. S	G. E. Leary & Son...	3	400	1½	350	400	Magothy	...	+4	5	...	Hard	...
35	Stillpond, 7 mi. S	Mrs. Janver	78	256	4	180	250	Magothy	...	-65	...	10	Hard	1904
36	Tolchester	Tolchester Beach Co...	30	60	1½	60	60	Matawan?	...	...	...	...	Hard	...

* Tapped 4 feet above creek, at which point it flows slightly.

THE SOILS OF KENT COUNTY

BY

JAY A. BONSTEEL

INTRODUCTORY.

Kent County lies entirely within the Coastal Plain and the various geologic formations which constitute the land mass of the county consist of unconsolidated gravels, sands, and clays. These different materials, though they have only passed through the first stages of rock formation, fall within the limits of the geologic definition of a rock, for they constitute an integral part of the earth.

The Eocene and Cretaceous sediments consist of greensand marls, in some instances containing fossil shells. The greensand is made up of the mineral glauconite and of medium to fine-grained quartz sand. The glauconite, being a silicate of potassium and iron, has a distinct value as a source of potash salts and for this reason it is frequently used as a fertilizer. It has been used in Kent County and several old marl pits are found along the Sassafras River. The weak action of this fertilizing material, and the fact that its value depends on the small amount of potash to be derived from it, has led to its abandonment in favor of commercial fertilizers which contain larger amounts of potash besides other plant foods.

The Miocene deposits, consisting largely of clay, are only sparingly represented by small areas in the southern part of the county and these are so covered by later deposits that they rarely influence the character of the soil.

The latest geologic formations found in Kent County are the ones which give rise to by far the larger part of the soils of the

county. They belong to the Pleistocene. Like all the other formations in the Coastal Plain they owe their origin to the deposition
of sediments over a tide water area. The materials consist of gravel
and sand arranged in layers or strata. Two well defined levels covered by Columbia deposits are found in Kent County. The higher
upland area, as described in the chapter on physical geography, consists of a mass of Cretaceous and Eocene material over which has
been laid down a layer of sand and gravel, coarser and thicker toward the northwest and gradually becoming thinner and composed
of finer particles toward the east and south. The greatest thickness
is found near the mouth of the Sassafras River between Coleman
and Chesapeake Bay, where this horizon reaches a total thickness
of about thirty feet. It thins rapidly until near Langford its total
thickness is only about twelve feet, while near Millington it is only
five or six feet thick.

Over this sand and gravel member is found a thin layer of mixed
boulders, gravel, and loam which forms the stony bands found along
steep slopes when it is exposed by stream erosion and mingles with
higher lying materials to form the Sassafras gravel loam on more
gently sloping areas.

The latest and highest lying material on the upland is a yellow
or reddish-yellow loam which forms the Sassafras loam and covers
the greater part of the county above 50 feet elevation. This material is also thicker toward the northwest, where it reaches a depth
of about fifteen feet, and thinner toward the southeast, where it is
absent from part of the area and its place is taken by the gravel
loam.

In the extreme southeastern part of the county east of Massey
and Millington the surface material is a very sandy loam, giving
rise to the Norfolk sand type of soil.

The lower foreland area, extending from tide water to about 50
feet elevation, forms the late stage of deposition in Kent County.
The latest Columbia strata are deposited over the inclined strata

FIG. 1.—VIEW OF A SAND SPIT AT THE MOUTH OF FAIRLEE CREEK.

FIG. 2.—VIEW SHOWING THE HARVESTING OF WHEAT.

of Cretaceous and Eocene age. The Columbia consists of a basal gravelly layer, covered by drab and blue clay, over a large part of the foreland portion of the county. Locally, as near Emory's wharf between Rock Hall and Eastern Neck Island and in the vicinity of. Worton Point,. the surface material is a medium or fine sand which gives rise to the Norfolk sand type of soil. The clay, on the other hand, gives rise to two main soil types. The poorly drained areas constitute extensive flat meadow lands, while the portions which have been subjected to a longer period of atmospheric action have become weathered out to a mottled clay loam typical of the Elkton clay soil.

It is easily seen that the geologic agencies, active in forming the sediments of which Kent County is built, have provided a considerable variety of materials from which soils have been derived either directly or indirectly. Thus the greensands of the Cretaceous and Eocene times, though only forming small areas along streams not adapted to agriculture, furnished sands for the construction of newer strata in Columbia time which do form portions of the arable land of the county. Thus the sandy southeastern portion of the county owes its character in part to the presence of these older sands, while the sandy areas of the forelands have been formed by an even larger deposit of Cretaceous, Eocene, and even older Columbia sands carried down by streams which were cutting into these older layers.

A large share of the Columbia sediments was derived from land areas outside of Kent County. The boulders and gravels of the Columbia are worn fragments of sandstone, shale, quartzite, granite, gabbro, and diabase, which correspond exactly to similar rocks still found in places along the Susquehanna River. The larger of these boulders could not have been carried to their present position by flowing water alone. Some of them are masses of rock of one or two tons weight and it is only the buoying power of floating ice that enables water to transport such coarse materials. Moreover every spring just such rocks are carried down the Susquehanna, borne upon the ice or frozen into the larger cakes which form along the stream margin in Pennsylvania and Maryland. The melting of these cakes drops the stones upon the bottom of the bay, just as the Columbia boulders were deposited in a former geologic period. Thus the presence of such large masses of rock indicates the existence of a land area whence they were derived; the presence of a Columbia stream, corresponding at least in part to the Susquehanna; and the prevalence of climatic conditions perhaps a little colder though not far different from those known at present.

The derivation of several distinct soil types from a single geo-. logic formation arises from the fact that the geologic classification of sedimentary rocks has for its basis the criterion of age, as determined by the character of life forms and by the position and succession of strata, while the soil classification depends upon the texture and structure of the particles composing any given soil. Thus a medium-grained sandy loam is adapted to the same crops, other things being equal, no matter whether it is of Columbia, Eocene, Cretaceous, or older geologic age. Similarly the character of marine sediments laid down during a single geologic age may vary considerably in texture and structure with different depths of water, with the character of material upon which the waves and streams are working, and with the presence or absence of organic life. In this manner a single geologic horizon may give rise to two or more soil types. When the added influence of the degree and kind of action of atmospheric agencies like frost, rain water, and organic acids in solution is considered, it becomes evident that the number of soil types to be found in a given region will usually exceed the number of geologic formations in the same area.

The present geologic activity in Kent County is similar to that over nearly all land areas and consists of the tearing away of sand, silt, and clay by waves and streams; the transportation of these materials along the coast and down the streams; and their deposition in areas of more quiet water. The remains of oysters and other shell fish buried in these sediments will furnish data for the age identification of these deposits when at some future time they shall come to form part of the land area.

THE SOIL TYPES

The soils of Kent County comprise the following distinct types, which will be summarized briefly.

1. *Sassafras loam.*—A yellow or brown loam about ten inches deep, underlaid by a heavier yellow loam subsoil. It occupies the greater part of the gently rolling upland and is well suited for general agricultural purposes.

2. *Sassafras gravel loam.*—Brown gravelly loam nine inches deep, underlaid by a red gravelly loam to thirty inches. This is in turn succeeded by red sand and gravel. It occupies sloping upland and produces corn, peaches, pears, and canning crops to good advantage.

3. *Susquehanna gravel.*—A loamy soil of about twelve inches depth, containing from 30 to 60 per cent of rounded gravel. It is usually underlaid by gravel beds. It is found near the tops of slopes, appearing only as narrow bands more or less continuous.

4. *Norfolk sand.*—A coarse sandy soil of from six to nine inches depth. The subsoil is a coarse yellow sand extending to a depth of three feet or more. The areas of this soil occupy the low terraces along river necks or small areas on the upland. It is a typical early truck soil.

5. *Elkton clay.*—A brown loam soil about nine inches deep, underlaid by a heavy mottled or gray clay loam. It occurs on the lowest terrace and is adapted to wheat and grass. The areas need extensive underdrainage.

Meadow.—This is in turn used to describe areas of low-lying, poorly drained, flat lands suited to grass and grazing. With proper underdrainage many of these areas are capable of producing good crops of grass and wheat.

THE SASSAFRAS LOAM.

The Sassafras loam covers a total area of over 130 square miles, lying wholly within the upland portion of Kent County. The soil is typically represented both in Kent and in the Coastal Plain portion of Cecil County, though it is by no means confined to these areas nor to the Eastern Shore of Maryland. It forms a portion of the widely extended sedimentary deposits of one of the latest geological formations along the Atlantic coast. It was deposited as a marine sediment during a geologically recent submersion and it partakes of the nature of materials usually laid down in waters of moderate depth. The layer or stratum of material from which this soil type is derived was deposited over tilted areas of a much older age. Its composition and the evidence of associated debris—chiefly boulders and blocks of stone scattered through the stratum—show that this soil material once formed a portion of higher-lying rocks and soils along the lower course of the Susquehanna River. The blocks of stone mentioned may be identified as once forming parts of ledges of sandstone, shale, conglomerate, quartzite, or masses of igneous rocks like the granite and gabbros found in Cecil and Harford counties. These rocks, and the finer particles which once covered them as soil in their original positions, have been carried by water and by floating ice to their present positions, where they are now rebuilt into a much newer geological formation. They have since been modified by the action of frost, rain, and organic agencies to form a valuable and highly productive soil.

In this first form these sediments constituted a nearly continuous sheet overlying the upland area; but stream drainage, as it became established, has cut into this material, added to slight original inequalities of surface, has relieved the monotonous level of the country and the soil now occupies the highest positions along interstream divides as well as along the main upland.

As a rule the surface of this formation in Kent County is slightly rolling and the areas possess sufficient irregularity of surface to allow of good natural drainage. In some instances small saucer-shaped depressions still exist unaffected by the general stream erosion, but short surface ditches or, better, a well-like drain down through the subsoil to underlying sandy layers will suffice to bring these wet places into good cultivation.

The soil proper consists of a fine brown loam which is often slightly sandy, especially in the eastern part of the county. It extends to an average depth of about nine inches and is underlaid by a typical yellow loam subsoil. The subsoil varies in thickness from about twenty inches to a maximum of five or six feet. It forms a supply reservoir capable of maintaining a large amount of soil moisture during the growing season, and it is as important a factor in the productivity of this soil type as is the soil proper. Underneath the true subsoil is usually found a layer of rather coarse gravel mixed with large-sized boulders and coarse sand, frequently cemented to a solid mass by the long-continued deposition of hydrated iron oxide. When in this state the gravel band is commonly known as hardpan in this region.

Below the gravel layer there is usually found a bed of medium to coarse red sand mixed with fine gravel and interspersed with seams and beds of gravels. Sometimes masses of clay are incorporated with this material. These lower-lying, coarser materials have little effect upon the higher upland areas of this soil type beyond furnishing a natural underdrainage, but where the higher lying surface becomes thinner as it descends toward stream beds the lower-lying, coarser material sometimes mingles with the finer-grained soil material sufficiently to produce a different soil condition.

The Sassafras loam is carefully cultivated over almost its entire extent, hence little if any of its original tree growth remains to indicate what the natural productivity of the soil brought forth.

The soil is well adapted to general farming. It lies between the limits of the heavy clay soils and the light sandy soils and is capable of producing a wide range of crops in generous amounts. It forms the typical corn and wheat soil of the county, producing wheat at a rate of from 15 to 20 bushels per acre—the quantity varying with the season and with the state of cultivation of different farms. Corn yields about 50 bushels per acre. Large orchards of Keiffer pears are found on this soil and while peach raising is not so largely followed now as formerly, many peach orchards both old and new are found on the Sassafras loam. The production of tomatoes, peas, and of other canning crops is also carried on upon this soil, while extensive asparagus beds are found in its area. Stock raising and dairying are followed and many flocks of sheep are to be found, chiefly upon this soil formation.

The diversity of interests already supported by this soil mark it as a highly valuable farming area for general purposes.

The Sassafras Gravel Loam.

In many instances where the slope from higher to lower levels is not steep enough to bring the heavy gravel band of the upland region to the surface as an outcrop, areas of decidedly gravelly soil are found. These owe their origin to the fact that the Sassafras loam is not so thickly developed over the areas as to cover in and obscure the underlying gravel completely, though enough of the finer material is present to constitute by far the larger part of the soil mass. Such areas are usually found on long, gentle slopes near or between the larger stream courses. Large tracts occur northwest of Millington and northwest of Chestertown, while smaller areas are found throughout the upland part of the county.

The surface of this soil type is generally sloping or rolling, and some of the smaller areas occur as bands along the gently sloping banks of smaller streams and near stream heads.

The soil consists of a brown, slightly sandy loam containing a scattering of gravel. This is underlaid by about two feet of heavy red or reddish-yellow loam, also containing gravel, which is in turn followed by red sand and gravel mixed with iron crust.

The less depth of heavy subsoil in this type and the consequent influence of the underlying sands and gravels are more important factors in differentiating it from the Sassafras loam than is the presence of the gravel in the soil. All the factors, however, combine to constitute a lighter soil type and to make it adaptable to other agricultural purposes.

Like the Sassafras loam this soil comprises lands, chiefly cleared, which have long been cultivated. The absence of natural tree growth precludes any conclusions drawn from natural conditions.

The Sassafras gravel loam approaches more nearly to a sticky corn-producing type than to a wheat-land, and it is also suited to the production of late truck crops like those used in the canning industry. Sugar corn, tomatoes, peas, and other crops produce well on similar soils and the climate of Kent County favors these crops. Nursery stock and small fruits can be raised on this type of soil, and while the wheat crop usually produces best on heavier soils a fair crop can be raised on the Sassafras gravel loam.

The Susquehanna Gravel.

Along the slope which separates the upland portion of Kent County from the foreland areas and along the steeper slopes down to stream areas the stony and gravelly layer underlying the Sassafras loam almost universally reaches the surface and its materials mingle with those of overlying and underlying formations. Thus a narrow band of steeply sloping stony soil is formed which makes a marked line of separation between other distinct soil types. Originally this layer of gravel could only have formed a narrow band of a width equal to the extent of the beveled edge reaching the surface on the slope; but long-continued freezing, thawing, rain washing, and the action of gravitation have spread the stone and gravel over wider areas and produced a stony slope soil.

This soil contains from 20 to fully 50 per cent of coarse gravel in certain places. The other finer material may be sandy, especially on slopes where the underlying sand formations reach the surface, or it may be composed of silt and clay washed down from the Sassafras loam.

The stony areas are frequently cultivated to the same crops as the other soils above and below, but they differ from them largely in ease of cultivation and in the varying degrees of productivity. Usually they are not sufficiently extended to warrant any special treatment or crop, though some of the slopes closely resemble soils devoted to vineyard interests in other localities.

It would not be possible to completely remove even the larger stones from these areas, as the supply from the gravel bands is almost inexhaustible and new crops of stone would work out into the soil so long as cultivation and atmospheric influences have access to this material.

The Norfolk Sand.

The Norfolk sand covers a total extent of nearly thirty square miles in Kent County. The largest single area of this soil type occurs in the southeastern part of the county. Here the surface of the land rises from near tide level along the Chester River to elevations exceeding 60 feet. The surface is gently rolling and quite generally forested. The higher elevations consist of rounded hills and hummocks of sandy soil, interspersed with hollows which are usually swampy and contain accumulations of partially decayed organic matter mixed with silt.

Along the shore of Chester River the lower lying land is quite generally sandy from near the water's edge up to 20 feet elevation.

In the foreland region of Kent County, beginning near Chestertown, there are found detached and scattered areas of this sandy soil, often comprising two or three square miles each. From Rock Hall southward to Eastern Neck Island this soil is also predominant, though composed of slightly finer material than elsewhere in the county. Near Worton Point and on the extreme end of Still Pond Neck this soil is again present in its coarse phase.

Other smaller areas of Norfolk sand are found over the upland, while the outcrops of the sandy underlying strata of Cretaceous, Eocene, and even of Pleistocene age in the deep stream cuts along the Sassafras River give rise to small areas of Norfolk sand.

The areas of this soil found along the forelands are usually slightly rolling or nearly flat, while those along the stream cuts are frequently very steeply inclined and consequently of little agricultural value.

The original sources of the sand entering into the composition of the Norfolk sand vary in different parts of the county. The greensands of Cretaceous and Eocene age consist of rounded quartz grains, glauconite, and some silt and clay. The weathering of outcrops of this material gives a sandy soil, usually found only along very deep stream cuts. But this same material when re-worked by streams and waves, transported to new localities and redeposited as a later sediment, also forms a soil which has the same agricultural values as along the weathered outcrops. In some instances it is possible to secure materials along the present shores from Cretaceous or Eocene outcrops, from the Pleistocene sandy stratum, and on the surface of the new foreland terraces, which differ from one another chiefly in the amount of the glauconite still present. Texturally they vary but slightly. Owing to this fact areas due to all these different causes have very similar crop values and are included in the same soil type.

The soil of the Norfolk sand consists of a medium to rather coarse sand with gravel also occurring in some areas. The soil is usually brown or reddish-brown, from the admixture of organic matter. It has a depth of about nine inches. The subsoil also consists of a medium sand, generally red or yellow, frequently containing sufficient silt to make it slightly adhesive.

The steeper sloping areas of Norfolk sand have not been entirely cleared and they are usually marked by a growth of chestnut and oak. The chestnut is found growing on this soil more frequently than on any other.

The Norfolk sand is a typical truck soil, although not actually used for such crops to any great extent in Kent County. Near Chestertown and near Worton Point truck and small fruits are being cultivated on this soil, but it is usually farmed in the regular rotation used in the county. The excellent facilities for transportation and the proximity of several large cities should lead to a more pronounced specialization of crops in this region and the Norfolk sand areas should be utilized as the best truck soil existing in Kent County.

THE ELKTON CLAY.

The Elkton clay occupies a total area of over twenty-five square miles in the foreland portion of Kent County. It usually lies between 15 and 40 feet elevation and its surface is very nearly level, or at most only gently sloping. The larger areas of the Elkton clay are found along the bay shore and on the necks which extend out into the Chester River. Only small areas of this type occur eastward from Chestertown in the southern part of the county and it is only represented by a single area on the Sassafras River just east of Shell Cross Wharf.

The materials forming this soil were deposited as a marine sediment during the Wicomico stage of the Columbia, and they have since been elevated to their present position above tide water. The low foreland area is largely made up of the same material, but it has not all proceeded to the same stage of soil formation. It will be noticed, with respect to the Elkton clays in Kent County, that all the areas lie in positions favorable to natural drainage; that is, they have the advantage either of considerable elevation above tide water or else of so lying that the slopes to natural drainage-ways are short and steep. It is due to this position and to the progress of natural underdrainage that most of these areas have been naturally brought to a more productive state than the surrounding meadow lands.

The first processes of soil formation, when any area of sediment becomes a part of the land, are those of drainage and of weathering. The rainfall must be disposed of and where the slope is sufficient streamways are formed which dispose of the surface waters. If the material is not too impervious a large part of the rain water percolates through it and finds an underground outlet to main drainage ways. The water passing underground carries various acids in solution and these aid in soil preparation.

The circulation of air also goes on, unless the soil pores are filled with water. So when air and water circulation are freely established various chemical and mechanical changes prepare the soil for crop production, but if they are interfered with these changes progress more slowly and the soil is considered wet, cold, and sour.

The materials constituting the meadow areas and those of Elkton clay are frequently the same, but the natural processes of soil formation have proceeded much farther in the latter case than in the former.

The Elkton clay is a yellow to brown silty loam soil, extending to a depth of about nine inches. This is underlaid by from twelve to thirty inches of mottled gray and yellow clay loam, which grades imperceptibly downward into a heavy dense drab clay. The drab clay was the original form of this material, but the circulation of air and water and of the solutions of various chemical compounds in the water has changed the upper portion of the clay while surface cultivation has changed the structure of the soil proper and mingled with it various amounts of organic matter.

The yellowing and mottling of the subsoil are due to the oxidation and deposition of iron salts held in the soil water and this process is still in progress. It has made the heavy plastic clay more loose and friable and this aids the underground circulation of soil water.

The growth, death, and decay of organic matter on the surface of the soil and the incorporation of this organic matter with the soil not only furnishes a temporary mulch for the retention of moisture within reach of growing plants but additional organic acids for the further preparation of the deep subsoils.

The natural growth over a large part of the Elkton clay included white oak, pitch pine, and sweet gum trees. Some areas still retain this growth while others, which have only been cleared in recent years, are not fully prepared for their best work in crop production.

The Elkton clay is more typically a wheat and grass land than any other soil type in the county. The soil and subsoil are sufficiently retentive of moisture to enable grain crops to maintain a steady growth, except during extremely dry seasons. The chief difficulty attending the cultivation of this soil is its tendency to form into clods and lumps. Wheat crops of from 30 to 35 bushels per acre are reported from different farms located on this soil type and good grass lands can be obtained. The hay is apt to be rather coarse and of only medium grade, but this is due fully as much to impure seed and lack of proper care as to any property of the soil.

Stock raising should be undertaken more extensively on this soil than it has been and the use of stable manures and lime may be profitably increased. Artificial underdrainage should be undertaken over considerable areas of the Elkton clay in order to facilitate the natural processes already under way.

THE MEADOW.

The meadow land in Kent County comprises areas of flat, poorly drained land best adapted to the production of grass and for pasturage. The meadows are not confined to soils of any one texture but are dependent for their characteristics rather on physiographic than on textural features.

The stream valleys are usually wet, poorly adapted to ordinary tillage, and are of greater value for grazing than for any other purpose. Certain parts of the upland portion of the county are so situated that the natural stream drainage has proved inadequate to prepare them fully for cultivation and they remain as forest areas. About one mile west of Massey an area of nearly three square miles still retains its meadow condition, owing to a lack of drainage, though the texture of the soil differs very little from the surrounding Sassafras loam. Two similar areas occur east of Chesterville, the northern one being above 60 feet in elevation and corresponding in texture to the Sassafras loam; the lower area sloping from 60 to 20 feet and resembling more nearly the Elkton clay. All three of these areas are so situated as to be easily reclaimed by artificial drainage. The lower-lying portions of the southeastern part of Kent County are also rather wet and fall within the meadow type, though a very little attention to drainage would fit them for the production of celery, cabbage, cauliflower, and late truck crops.

By far the largest meadow areas are found in the lowland division of the county, these areas usually lying between sea level and an elevation of 20 feet. They owe their present condition chiefly to lack of drainage.

The foreland portion of the county is the youngest part geologically and drainage systems are not yet completely established. As a result those areas which lie near water level are saturated nearly to the surface and the meadow condition is the only one possible.

The natural growth on all of the meadow areas consists of willow oak, sweet gum, and other water-loving forms. The main forest areas of the county are found on the meadow areas, though they are not all of them forested. Recent removal of forests has thrown some of the foreland areas into cultivation and wheat and grass are produced to fair advantage. The production of corn is not successful for in wet seasons planting is usually prevented until late on account of the water-soaked condition of the ground, and in time of drought the surface bakes to such an extent that growth is interfered with and the crop becomes yellow and backward. This yellowing is known locally as "Frenching." Underdrainage, to permit better circulation of the air, and frequent shallow surface cultivation, to form a soil mulch, would help to prevent this baking process.

The soil of the lowland meadows consists of a gray loam having a depth of about eight inches. The subsoil is a blue or gray clay loam which is very heavy and plastic when wet, but on exposure to the air usually bakes to a hard surface. The clayey subsoil contains considerable silt.

The meadow lands of Kent County may be reclaimed by underdrainage and added to the grain producing areas of the county. The upland meadows are so situated that drainage ditches may be cut to the heads of existing streams with laterals ramifying over the areas. Then local underdrains should be provided for each field. The only question involved is that of the comparison of the expense with the results to be obtained. The lowland meadows in some cases lie too near to tide level to be reclaimed easily, but many of the areas now grown up to sweet gum and willow oak could be made to produce wheat and grass if properly drained.

THE SWAMPS.

The swamp lands of Kent County fall into two classes: the salt marshes and fresh water marshes. The salt marshes occupy positions along the estuaries and are subject to inundation by the highest tides, while the fresh water marshes are usually formed along the upland streams where the surface slope is insufficient to carry off all the surface water. The salt marshes comprise by far the larger area. Neither type is at present of any great agricultural value. When the value of lands in the East becomes greater the tide can be excluded from the salt marshes by diking and artificial drainage will obliterate the fresh water marshes, but so much other land remains in the East either in forest or in a low state of cultivation that the marsh areas are apt to play but a small part in agricultural operations for many years to come.

THE AGRICULTURAL CONDITIONS

Kent County has been an agricultural community from the time of its early settlement to the present day. In the earlier times the county was divided into large manorial estates and later subdivided into smaller farms. Some of the farms have remained in the possession of single families for two hundred years. The effect of this long tenure is evident in the general prevalence of substantial farm buildings and in the state of cultivation to which a very large proportion of the land has been brought. Substantial houses are to be found in all parts of the county, each forming the center of a group of farm buildings. The boundary lines and roads are marked by osage hedges, and long avenues of trees leading from the main highways to the residences are frequently found.

The early crops were largely confined to the grains, while within recent years the cultivation of truck and canning crops has been introduced. The greatest change of recent years, however, began with the rise of the peach industry. Thousands of acres were devoted to peach orchards and a full crop and fair prices brought excellent returns. For many years the peach crop was maintained, but the opening of new areas to the cultivation of the fruit affected the markets and as the orchards grew older they became more subject to various diseases, in spite of every care; and at present the acreage devoted to peaches is decreasing rather than increasing.

The Keiffer pear has been introduced along with other varieties and proves a wonderful producer. The pears are sold to local canning companies at varying prices and even at the lowest price some profit is derived. Tomatoes are raised extensively as a canning crop and usually yield fair returns. Asparagus beds are found on many farms, while small fruits are being cultivated to a limited extent. The areas of Norfolk sand found in the county are well adapted to the production of truck; and small fruits such as strawberries, raspberries, blackberries, currants, and grapes should be added to the list of Kent County products.

Dairying, stock raising, and sheep raising are other farm industries of the county. Several creameries manufacture the milk. The dairy industry should be made to supplement the canning industry. Sweet corn can be produced in Kent County for canning purposes. The forage crop remains and may be cured and stored for dry feeding or, better, may be shredded and stored in silos for green feeding. The advantages to be derived from the cash return from the canning factory and the creamery are not the only benefits to be obtained from this practice. The item of farm expense annually charged to the fertilizer bill may be very largely eliminated by the production of increased amounts of stable manure.

Every bushel of grain sold from a farm removes absolutely beyond recall so much plant food. The store of such food in the soil must be replenished and commercial fertilizers are resorted to. On the other hand the dairying and corn producing rotation give immediate cash return, with little or no drain upon the supply of plant food. Moreover the item of transportation charges is also reduced. The nearness of such markets for dairy products as are furnished by Washington, Baltimore, and Philadelphia should awaken the community to the desirability of increased dairying along the most modern lines of development.

TRANSPORTATION.

Kent County is well situated with respect to transportation facilities both for internal communication and for egress to the centers of commerce and trade along the Atlantic seaboard. The county is bounded by over eighty miles of coast line. The head of navigation on both the Sassafras and Chester rivers is not reached until near the Delaware line, and the entire western limit of the county is formed by Chesapeake Bay.

Several steamboat lines carry freight and passengers to Baltimore and Philadelphia and during the grain and fruit seasons extra freight steamers are provided. Ice interferes with navigation only during periods of excessive cold.

In addition to the opportunities for navigation two railroads cross the county. One has its terminus at Chestertown and at Clayton, Delaware. The other connects Centerville, Queen Anne's County, with the trunk lines farther north. It enters Kent County at Millington and crosses the Delaware line at Golts. The railroads cross each other at Massey and together furnish rail communication with trunk lines. In addition, various passenger and freight automobile lines have been inaugurated in recent years.

THE CLIMATE OF KENT COUNTY

BY

ROSCOE NUNN*

INTRODUCTORY.

Kent County lies wholly within the region known as the Coastal Plain Province and its topography is simple. The county is, roughly, crescent-shape, its greatest length being along a nearly east-west line. Its southernmost margin is in latitude N. 39° 1' and its northernmost limits in latitude N. 39° 23', while in longitude it extends from W. 75° 46' to W. 76° 17'. Its area is about 281 square miles. Bordered as it is on its comparatively long western side by the Chesapeake Bay and on its north and south sides by the broad estuaries of the Sassafras and the Chester rivers, respectively, a large portion of the county is affected by the tide waters of the Chesapeake. The drainage is about equally divided between the watersheds of the Sassafras River on the north and the Chester River on the south, both of which flow into Chesapeake Bay.

In recent years Kent County has been well represented in our climatological investigations and studies. A fairly good view of the climatic conditions is now presented by these records. Acknowledgment of the valuable service rendered by the coöperative observers is made with pleasure at this time. While the instrumental equipment and the supervision of the work were furnished by the United States Weather Bureau, in coöperation with the Maryland State Weather Service, the results would not have been possible except through the conscientious work of the coöperative observers.

* The paper is prepared by direction of Dr. Edward B. Mathews, Director, Maryland State Weather Service. Credit is due Mr. Joseph Bily, Jr., and other assistants for valuable aid.

A list of the Kent County climatological stations follows:

CLIMATOLOGICAL STATIONS.

Climatological records are available from the following stations:

BETTERTON.—Elevation, 80 feet; extreme northwestern portion of Kent County; on south shore of Sassafras River, at mouth; about 2 miles due east from Chesapeake Bay. Observations from March to July, 1898, by Mr. Edward E. Carey, under Maryland State Weather Service.

CHESTERTOWN.—Elevation, 85 feet; about half-way between extreme southwestern and southeastern corners of Kent County; about 25 miles, on north shore, from mouth of Chester River, which is boundary between Kent and Queen Anne's Counties. Observations from June, 1855, to July, 1864, were under the auspices of the Smithsonian Institution, and were made at Washington College, by Prof. J. R. Dutton and Prof. F. L. Bardeen. There is a private record from 1880 to 1895. Under the Maryland State Weather Service and the U. S. Weather Bureau, observations from November, 1893, to March, 1910, were made by Hon. Marion De Kalb Smith; and under the U. S. Weather Bureau from April, 1910, to December, 1913, by Mr. M. W. Thomas.

COLEMAN.—Elevation, 80 feet; extreme northwestern portion of Kent County; about 3 miles south of Betterton and the Sassafras River and about 3 miles east of Chesapeake Bay. Observations under the U. S. Weather Bureau were made from February, 1898, to September, 1914, by Mr. James Sheppard Harris; from January to May, 1915, by Mr. Carson W. Harris; and from September, 1916, to December, 1925, by Mr. Walter B. Harris.

GALENA.—Elevation, 60 feet; northeastern portion of Kent County; about 2 miles south of Sassafras River and about 6 miles west of the eastern boundary of the County. Observations were made under the U. S. Weather Bureau from September, 1888, to June, 1890, by Mr. Henry Parr.

MILLINGTON.—Elevation, 27 feet; extreme southeastern portion of Kent County; on Pennsylvania Railroad; about ¾ mile north of Chester River and about 4 miles west of eastern boundary of the County. Observations under the U. S. Weather Bureau were made from October, 1898 to May, 1906, by Mr. J. S. Barwick; from October, 1906, to January, 1912, by Mr. James E. Higman; and from February, 1912, to December, 1925, by Mr. Henry L. Higman.

ROCK HALL.—Elevation, shore station (No. 1), 20 feet, inland station (No. 2), 25 feet; extreme southwestern portion of Kent County; on Rock Hall Creek on Chesapeake Bay; about 9 miles north of extreme southwestern corner of the County. Observations were made under the Maryland State Weather Service: At station No. 1, in March, 1898, by Mr. Charles R. Kerr, and from April, 1898, to May, 1900, by Mr. Charles Nathan Satterfield; at station No. 2, from February, 1898, to February, 1902, by Mr. Isaac Lassell Leary. Under the U. S. Weather Bureau observations were made at station No. 2 from September, 1919, to February, 1920, by Mr. George R. S. Downey, and from July, 1921, to December, 1925, by Mr. Charles Judefind.

DATA AVAILABLE

It will be seen that weather records began in Kent County as early as 1855 and that records have been kept continuously from November, 1893, to the present time, although no one of the stations has an unbroken record for quite so long a period. Under the auspices of the Smithsonian Institution, the first observations were begun in June, 1855, by Prof. J. R. Dutton, of Washington College, Chestertown. This early series continued, with considerable interruptions, until July, 1864.

From 1894 to 1913, inclusive, the Chestertown record is only slightly broken. For the years 1898 and 1899 and from 1902 to 1925, inclusive, the Coleman record has but few interruptions. From 1899 to 1925, inclusive, the Millington record has only one or two slight breaks. A good record was kept at Rock Hall from September, 1919, to date, and there are also short series of records for this station at intervals prior to 1919. There are fragmentary records for Galena and Betterton.

These records are sufficient to give fairly reliable monthly averages of temperatures, rainfall, and snowfall, and the average time of last killing frost in spring and the first in autumn; also, the frequency of some of the important climatic elements, such as extremes of temperature, excessive precipitation, drought, and thunderstorms. However, there is no doubt but that longer records will depict more definitely the characteristics of the climate.

CLIMATIC FEATURES

It is interesting to note that the early records at Chestertown show the coldest month in the entire record for Kent County. This very cold month was January, 1856, when the monthly mean temperature was 21.4 degrees. Records at Baltimore and other stations bear out the Chestertown record and show that January, 1856, was one of the coldest months of the last century.

The tables require little explanation. From an inspection of them it may readily be seen how temperature and rainfall vary from month to month, and what ranges may occur in the records of any month over a period of years. They show how cold or how warm and how wet or how dry any month has been within the period of history here given; also, what extremes of temperature and precipitation may be expected, judging from the past.

The long western shores of Kent County enjoy a slight amelioration of temperature conditions, as compared with the eastern or interior portions of the county. This, in a slight way, illustrates the well known fact that a body of water of any considerable proportions on the windward of a land area modifies the temperature extremes that otherwise would be experienced. It is shown in the length of the growing, or frost-free, season, which is about ten days

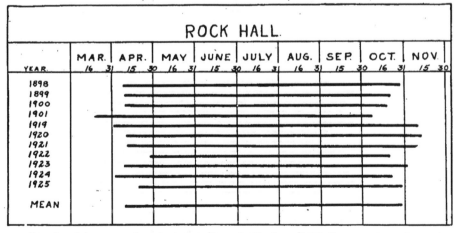

Fig. 2.—Diagrams showing variations in the length of the growing season at Coleman and Rock Hall.

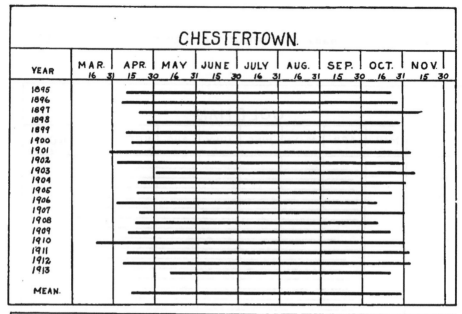

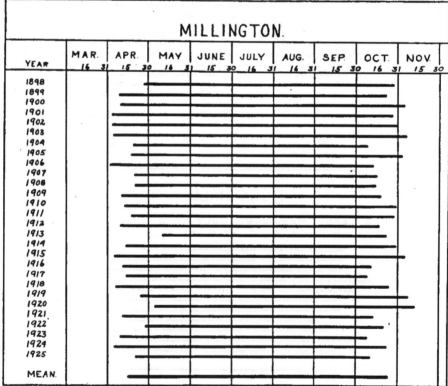

Fig. 3.—Diagrams showing variations in the length of the growing season at Chestertown and Millington.

longer on the western border of the county than on the east. For example: the average date of last killing frost in spring at Rock Hall is April 9; at Millington, April 15; and the average date of first killing frost in autumn at Rock Hall is October 28, while at Millington it comes on October 23.

Certain facts not brought out by the tables, but which are discoverable from the detailed records of the observers, are given below. These facts, representing the county as a whole, are gathered from the combined or average records for Chestertown, Coleman, Millington, and Rock Hall.

Temperature, frequency of certain extremes.—The average number of days in the year with maximum temperature as high as 90° or above is 23; with minimum temperature as low as 32° or below, 98; with minimum as low as 14° or below, 11. Temperatures of zero, or lower, rarely occur. Millington records zero or lower in only nine years of the last twenty-eight; Rock Hall had zero or lower in only one year of the last nine.

Length of growing season.—The average date of the last killing frost (freezing temperature) in spring is April 12; average date of the first in autumn, October 26. This gives an average of 197 days for the growing season. The extremes of killing frost dates are as follows: Earliest date of last in spring, March 19; latest date of last in spring, May 12. Earliest date of first in autumn, October 8; latest date of first in autumn, November 13.

Sunshine and cloudiness.—The average number of days in the year with cloudy sky is 119; partly cloudy, 132; clear, 114.

Humidity.—No records of relative humidity are available for Kent County. However, from records kept at neighboring regular Weather Bureau stations, especially Baltimore, it is apparent that the average relative humidity in Kent County is about 70 per cent. This is slightly lower than on the immediate Atlantic coast and slightly higher than in the Piedmont and Appalachian mountain regions of Maryland.

Excessive rainfall.—Rainfall of 2.50 inches or more within a 24-hour period is called excessive. Such excessive falls have occurred 25 times at Coleman and 19 times at Millington during the last twenty-eight years and at Rock Hall 6 times in the last nine years. This gives an average of less than one occurrence a year. These excessive rains fall principally in the month of July and August, as thunder showers, but almost as frequently in September, mostly in connection with Atlantic coast storms. The greatest 24-hour rainfall recorded in Kent County was 5.45 inches, in July, 1901. Amounts greater than 4.00 inches within 24 hours are very rare.

Thunderstorms.—In the late autumn and during the winter thunderstorms seldom occur, but a few have been recorded even in mid-winter. On an average in this region thunderstorms occur on about 36 days in the year. The average number for May is 4; June, 7; July, 9; August, 7; September, 3. Thunderstorms occur nearly three times as often in portions of Florida and twice as often in southern Georgia, Alabama, and the middle Gulf region.

Tornadoes.—These, the most violent of all local storms, are unknown in Kent County.

Prevailing winds.—The wind comes from a westerly direction (northwest and southwest principally) during most of the winter and early spring seasons and from the south and southwest in summer and early autumn. There is a considerable percentage of winds from the north. Easterly winds are the most infrequent.

Droughts.—During the crop growing season droughts causing serious damage have been recorded only two or three times in the last thirty years. It will be noted from the precipitation tables that July and August have greater average rainfall than the other months of the year and that June is about the third month for plentiful rainfall. Droughts occur more often in the autumn, but at that season are seldom injurious from any standpoint.

CONCLUSIONS

Having made comparisons * of the climatic data for Kent County with the data for many other regions, the writer is prepared to say that this county is favored climatically on account of its geographical position and its insular setting. The climate is free from great extremes. It is conducive to comfortable living the year round and is favorable for the industrious occupations of the people. The winters are mild for the latitude, with light snowfall and a large percentage of sunshine. The summers are warm, but with seldom such prolonged spells of hot weather as occur farther west and south. Precipitation is ample at all seasons, but rarely excessive. The growing season, or period between the last killing frost in spring and the first in autumn, is long and favorable for a diversified agriculture.

* It should be borne in mind that, to get a correct and clear understanding of the climate of any region or locality, it is necessary to compare climatic statistics. It is quite desirable for one to be familiar with the numerical climatic data for his own locality in order to compare climatic data for other places understandingly. While it is impracticable to give in this paper comparative data for other regions, such data may be procured for almost any locality of the United States by writing to the U. S. Weather Bureau, Washington, D. C., or to the nearest Weather Bureau station.

TABLE I.*

MONTHLY AND ANNUAL MEAN TEMPERATURES AT BETTERTON.

Year	Jan.	Feb.	March	April	May	June	July	Aug.	Sept.	Oct.	Nov.	Dec.	Annual
1898	....	....	48.9	50.8	63.3	73.0	79.1	....	....	....	....	....	....

AT CHESTERTOWN.

Year	Jan.	Feb.	March	April	May	June	July	Aug.	Sept.	Oct.	Nov.	Dec.	Annual
1855	....	....	....	....	....	75.3	77.9	72.6	....	52.0	45.7	35.6	....
1856	21.4	24.7	....	....	....	....	....	....	....	....	....	....	....
1857	....	....	....	....	....	....	74.4	....	....	....	....	....	....
1858	39.0	28.7	40.9	51.7	58.8	75.2	78.4	74.8	66.6	57.7	41.0	38.4	54.3
1859	33.9	36.2	46.9	48.6	62.8	69.3	75.0	73.6	66.4	50.2	46.5	33.4	53.6
1860	....	....	....	....	....	....	....	....	....	....	....	....	....
1861	31.2	40.1	43.6	52.6	59.7	72.9	75.0	76.0	69.9	60.2	44.9	37.2	55.2
1862	34.2	33.8	40.7	49.4	62.5	68.8	74.8	75.9	.1	58.8	47.7	35.5	54.0
1863	....	....	....	....	64.4	69.5	76.6	....	....	54.8	47.8	36.0	....
1864	34.2	35.9	39.0	50.9	67.4	71.0	77.5	....	....	....	....	....	....
1893	....	....	....	....	....	....	....	....	....	....	46.8	34.9	....
1894	37.8	33.8	48.8	53.0	68.8	71.8	76.4	71.8	69.8	56.8	44.	36.4	55.8
1895	30.	23.	49.	51.9	.4	68.	72.0	74.	70.0	58.	45.	8.6	2.
1896	31.	4.	7.	44.6	.6	.	6.0	75.	67.0	.	50.	3.4	4.
1897	30.	5.	3.	1.8	.3	.	5.2	71.	65.9	.	44.	36.5	3.
1898	34.	3.	7.	9.2	.6	.	7.6	6.	69.6	.	43.	35.6	5.
1899	31.	6.	1.	1.8	.0	.	6.2	4.	66.0	5	.	44.	47.2	3.
1900	36.	34.	39.	3.0	.2	.	8.4	7.	72.2	.	49.	6.0	5.
1901	32.	30.	43.	0.1	.0	.	9.0	5.	67.4	.	39.	3.4	3.
1902	30.	28.	45.	1.6	.2	.	5.6	2.	66.3	.	50.	3.8	3.
1903	2.	35.	8	2.3	.2	.	5.2	1.	66.7	.	41.	1.6	3.
1904	6.	26.	1	9.2	.4	.	3.6	2.	66.3	.	41.	8.6	1.
1905	8.	24.	2	1.8	.0	.	5.5	3.	67.2	5	.	43.	8.0	52.9
1906	9.	34.	8	3.5	.1	.	4.8	6.	70.6	.	45.	6.4	55.1
1907	5.	27.	5	7.0	.4	.	4.7	1.	68.6	.	45.	8.4	52.5
1908	3.	30.	5	4.8	.2	.	8.2	2.	65.0	.	45.	6.9	54.8
1909	5.	43.	0	3.6	.3	.	3.6	2.	66.5	.	49.	2.2	5 .6
1910	2.	34.	8.	6.8	.2	.	7.6	3.	70.1	.	42.	0.2	5 .8
1911	6.	34.	1.	9.6	.2	.	8.4	5.	70.2	.	44.	1.8	5 .0
1912	6.	31.	1.	5.6	.2	.	5.7	3.	70.2	.	47.	7.7	5 .6
1913	3.	36.	9.	5.5	.6	.	7.0	4.	69.2	.	49.	1.2	5 .5
Av.	33.0	32.2	43.1	52.0	63.2	71.0	76.1	73.9	68.1	56.3	45.4	35.7	54.1

AT COLEMAN.

Year	Jan.	Feb.	March	April	May	June	July	Aug.	Sept.	Oct.	Nov.	Dec.	Annual
1898	35.0	34.0	48.3	52.0	62.6	72.7	78.6	76.1	70.7	58.5	43.7	35.9	55.7
1899	32.6	26.6	40.7	52.8	63.4	74.6	77.4	74.3	68.7	57.8	47.0	37.0	54.4
1900	....	....	....	....	....	....	....	....	....	....	....	....	....
1901	....	....	....	....	....	....	....	....	....	....	....	....	....
1902	30.	28.6	46.	52.	68.8	71.0	76.	72.	66.	57.0	5 .	34.	5 .
1903	38.	36.8	50.	56.	6	.6	67.4	76.	76.	68.	58.2	4 .	32.	5 .
1904	.	26.8	41.		6	.2	70.9	75.	.	68.	55.0	4 .	0.	5 .
1905	29.	25.2	43.	5	.	6	.6	71.6	76.	.	69.	8.	4 .	8.	5 .
1906	39.	35.0	38.	.	6	.8	73.4	75.	.	72.	7.	4 .	6.	5 .
1907	38.	28.2	45.	.	5	.8	66.2	76.	.	70.	3.	4 .	9.	5 .
1908	.	30.0	46.	5	6	.4	72.8	79.	.	67.	0.	4 .	6.	5 .
1909	.	43.6	41.		6	.8	74.2	75.	.	68.	5.	4 .	2.	5 .
1910	.	34.0	48.		6	.4	69.0	77.	.	71.	0.	4 .	0.	5 .
1911	.	35.3	41.			.1	71.8	78.	.	69.	7.	4 .	0.	5 .

* Figures in italic denote interpolated data.

TABLE I.—Continued

Year	Jan.	Feb.	March	April	May	June	July	Aug.	Sept.	Oct.	Nov.	Dec.	Annual
1912	25.2	30.6	40.6	54.0	64.4	69.8	75.7	73.8	70.3	60.6	48.2	39.4	54.4
1913	32.8	35.8	48.6	54.8	64.6	72.6	77.4	74.8	68.8	68.8	47.8	39.8	57.8
1914	6.	9.	9.	.	6.	.	5.8	6.	8.	.	44.	3.	4.
1915	6.	8.	8.	.	61.	.	76.0	73.	70.	.	44.	34.	54.
1916	38.	32.	36.	.	65.	.	76.8	75.	66.	.	47.	34.	54.
1917	33.	31.	41.	.	58.	.8	76.	75.	65.	.	43.	88.	53.
1918	31.	33.	45.	.	68.	.	6.	7.	6.	.	7.	1.	55.
1919	7.	6.	6.	.	4.	.	8.	3.	0.	.	7.	1.	56.
1920	7.	1.	3.	.	0.	.	5.	4.	0.	.	6.	9.	54.
1921	6.	8.	4.	.	2.	.	9.	4.	3.	.	8.	6.	57.
1922	1.	7.	4.	.	6.	.	6.	4.	0.	.	8.	6.	56.
1923	4.	0.	3.	.	3.	.	5.	4.	0.	.	5.	3.	55.
1924	5.	3.	1.	.	9.	.	5.	4.	5.	.	6.	5.	53.
1925	0.	0.	5.	.	1.	.	6.	3.	3.	.	4.	6.	55.
Av.	33.3	33.2	43.9	53.4	63.6	71.9	76.6	74.4	69.2	58.3	46.2	35.9	55.0

AT GALENA.

Year	Jan.	Feb.	March	April	May	June	July	Aug.	Sept.	Oct.	Nov.	Dec.	Annual
1888	....	....	....	....	....	....	...	..	65.6	51.1	46.7	35.2	....
1889	36.5	29.6	41.8	54.6	65.2	73.8	75.8	72.6	65.5	52.7	47.5	43.6	54.9
1890	42.6	41.8	40.1	54.2	64.4	75.2	....	....	....	....	....	....	....

AT MILLINGTON.

Year	Jan.	Feb.	March	April	May	June	July	Aug.	Sept.	Oct.	Nov.	Dec.	Annual
1898	....	....	....	....	....	....	....	....	....	56.5	45.2	38.5	....
1899	35.3	28.0	44.8	53.	63.	73.	75.1	73.2	66.8	68.6	47.6	38.8	54.8
1900	37.5	35.1	40.				36.3	75.9	81.	.8	50.0	.	56.8
1901	35.4	29.0	44.				0.4	7.3	9.	.8	40.6	.	54.
1902	30.8	29.0	46.				7.4	3.4	7.	.2	51.4	.	55.
1903	33.2	38.5	51.				6.7	2.4	7.	.6	42.2	.	54.
1904	26.6	27.5	41.				5.0	4.6	9.	.9	42.0	.	52.
1905	29.3	25.6	46.				7.6	4.6	9.	.0	43.8	.	54.
1906	40.0	35.3	39.				75.0	76.0	72.	.6	45.0	.	55.
1907	35.5	27.4	46.				75.5	72.4	69.	.0	5.0	.	53.
1908	34.3	31.0	46.				78.4	71.9	66.	.9	5.4	.	55.
1909	36.0	42.8	40.				3.8	1.0	7.	.5	8.8	.	54.
1910	32.0	35.0	49.				6.8	2.8	0.	.4	1.5	.	54.
1911	36.9	35.2	41.				8.2	5.7	0.	.7	3.6	.	55.
1912	24.9	30.2	40.				5.2	2.8	9.	.4	6.7	.	53.
1913	42.5	35.1	48.				6.6	3.4	7.	.8	6.8	.	56.
1914	36.8	30.1	37.				5.4	6.0	5.	.0	4.7	.	54.
1915	35.9	38.5	38.				6.4	3.6	0.	.4	4.2	.	54.
1916	38.8	32.8	36.				6.4	5.9	6.	.8	6.2	.	54.
1917	34.0	32.1	42.				6.0	4.4	2.	.4	1.5	.	52.
1918	22.6	33.0	45.				4.0	6.1	3.	.3	4.9	.	54.
1919	36.2	35.8	47.				5.8	2.4	7.	.4	6.2	.	55.
1920	27.3	31.1	43.				3.6	3.9	7.	.4	5.2	.	53.
1921	36.0	38.0	55.				8.6	1.6	72.	.6	7.8	.	57.
1922	30.9	37.4	45.				5.2	2.6	68.	.6	6.8	.	55.
1923	35.5	30.6	44.				4.8	4.0	69.	.2	4.2	.	55.
1924	35.3	33.0	42.				4.4	4.6	4.	.2	5.3	.	5.
1925	31.0	42.3	45.				5.6	2.6	2.	.4	4.2	.	5.
Av.	33.7	33.3	44.0	53.5	63.2	71.3	76.1	73.9	68.4	57.3	45.2	35.8	54.6

AT ROCK HALL (No. 1)—(Shore Station).

Year	Jan.	Feb.	March	April	May	June	July	Aug.	Sept.	Oct.	Nov.	Dec.	Annual
1898	....	....	48.8	....	63.5	72.7	78.6	74.4	72.1	60.2	45.0	35.9	....
1899	34.8	27.7	42.2	....	....	74.6	77.4	74.4	....	....	....	....	....
1900	....	....	....	53.1	64.4	....	....	....	....	....	....	....	....

10

THE CLIMATÉ OF KENT COUNTY

TABLE I.—Continued
AT ROCK HALL (NO. 2).

Year	Jan.	Feb.	March	April	May	June	July	Aug.	Sept.	Oct.	Nov.	Dec.	Annual
1898	....	....	48.2	50.1	63.2	71.7	77.8	76.8	70.2	58.8	44.0	36.1	....
1899	33.0	36.0	44.9	52.9	63.6	73.9	79.	74.0	62.0	67.6	45.9	37.6	53.9
1900	36.6	.	.4	52.	.	.1	7.	9.	7.	2.	49.4	6.	56.
1901	.	.	.4	50._	.	.5	7.	6.	6.	5.	41.0	3.	53.
1902	.	.	....	....	.	.	....	.	.	.	....	.	....
1919	....	....	....	....	....	....	....	....	68.6	66.6	47.6	56.8	....
1920	27.6	32.6	44.6	53.6	59.6	71.6	75.6	74.6	70.	.	46.	.8	54.8
1921	37.	36.	45.	59.	66.	76.	79.	.	72.	.	48.	.2	57.9
1922	32.	.	4.	4.	.	.	76.	.	70.	.	48.	.8	56.4
1923	36.	.	4.	2.	.	.	75.	.	70.	.	5.	.2	55.7
1924	5.	.	1.	0.	.	.	4.	.	64.	.	6.	.2	53.7
1925	1.	.	5.	5.	.	.	5.	.	72.	.	3.	.4	55.3
Av.	33.3	33.5	45.0	53.1	62.1	73.2	76.8	75.0	69.3	58.5	46.0	36.9	55.2

TABLE II.
HIGHEST TEMPERATURES AT BETTERTON.

	Jan.	Feb.	March	April	May	June	July	Aug.	Sept.	Oct.	Nov.	Dec.	Annual
1898	....	....	76	78	90	96	101	....	....	....	....	....	101

AT CHESTERTOWN.

	Jan.	Feb.	March	April	May	June	July	Aug.	Sept.	Oct.	Nov.	Dec.	Annual
1855	....	....	....	....	....	....	....	....	....	....	....	....	93
1858	....	....	....	....	....	....	....	....	....	....	....	....	95
1859	....	....	....	....	....	....	....	....	....	....	....	....	95
1861	....	....	....	....	....	....	....	....	....	....	....	....	96
1893	....	....	....	....	....	....	....	....	....	....	60	63	....
1894	58	59	....	81	84	94	93	91	90	80	....	56	94
1895	57	58	66	81	92	93	90	....	....	74	75	65	93
1896	56	60	67	87	89	88	90	93	90	71	71	58	93
1897	63	56	72	81	79	90	88	85	90	81	64	....	90
1898	....	61	79	75	88	91	90	90	90	83	67	62	97
1899	56	56	69	78	87	94	91	91	89	78	64	67	94
1900	....	....	....	....	88	91	96	.	89	83	73	57	....
1901	58	49	70	77	79	....	100	88	86	76	63	74	100
1902	50	58	71	82	86	90	95	87	85	76	72	59	95
1903	54	67	71	83	87	87	92	91	84	81	68	57	92
1904	57	59	68	77	87	92	91	87	87	79	61	60	92
1905	58	46	75	78	85	92	93	88	84	80	64	58	93
1906	70	59	60	81	88	90	90	90	88	75	66	64	90
1907	66	50	82	77	80	88	88	88	85	73	63	62	88
1908	57	63	76	81	87	92	96	90	83	82	66	66	96
1909	58	71	71	81	86	92	92	92	85	76	72	56	92
1910	56	68	78	84	87	91	92	91	92	83	65	58	92
1911	59	63	69	77	94	95	98	94	86	76	77	63	98
1912	56	65	70	78	87	90	94	91	94	84	74	66	94
1913	63	66	75	80	89	96	97	95	92	78	76	61	97

AT COLEMAN.

	Jan.	Feb.	March	April	May	June	July	Aug.	Sept.	Oct.	Nov.	Dec.	Annual
1898	....	....	73	79	91	95	102	93	95	86	66	63	102
1899	55	58	68	80	89	97	98	95	90	79	69	65	98
1900	60	66	67	86	93	91	....	....	....	....	....	....	....
1901	....	....	....	....	.	98	104	....	....	....	....	....	104
1902	....	60	72	85	90	....	....	....	....	....	76	59	....
1903	53	67	73	88	92	87	97	95	91	84	74	53	97
1904	57	59	71	81	90	95	96	92	94	85	64	60	96

TABLE II.—Continued

Year	Jan.	Feb.	March	April	May	June	July	Aug.	Sept.	Oct.	Nov.	Dec.	Annual
1905	63	46	80	81	86	93	96	91	89	87	67	58	96
1906	72	61	60	84	92	92	92	93	92	79	67	68	93
1907	71	49	87	80	82	89	92	94	90	75	64	64	94
1908	....	....	....	85	88	96	99	95	85	86	71	69	99
1909	57	69	71	83	89	95	96	99	88	80	75	60	99
1910	....	....	....	83	85	92	93	92	95	86	64	55	95
1911	58	64	69	75	94	96	99	95	89	78	69	64	99
1912	55	57	70	....	85	89	95	94	94	85	75	67	95
1913	65	67	74	79	90	97	97	97	92	76	75	60	97
1914	69	57	72	80	93	98	100	98	91	....	....	....	100
1915	63	64	58	90	....	....	....	....	....	....	....	....	....
1916	....	....	....	....	....	....	....	....	....	86	73	67	....
1917	52	61	74	81	87	94	98	96	85	78	67	48	98
1918	58	60	76	78	90	94	96	106	86	80	69	65	106
1919	60	61	74	73	89	....	94	91	94	90	72	65	94
1920	50	50	74	81	83	95	91	91	87	85	70	68	95
1921	59	68	85	83	86	96	95	94	94	79	76	65	96
1922	55	70	78	87	87	92	94	91	93	92	72	56	94
1923	61	52	78	82	89	98	98	95	88	80	62	64	98
1924	63	51	72	78	86	96	97	100	95	80	73	67	100
1925	49	66	73	84	96	100	100	93	93	76	68	57	100

AT GALENA.

Year	Jan.	Feb.	March	April	May	June	July	Aug.	Sept.	Oct.	Nov.	Dec.	Annual
1888	....	....	....	....	....	....	....	....	....	62	....	61	....
1889	57	40	....	....	....	....	....	....	....	....	....	....	....

AT MILLINGTON.

Year	Jan.	Feb.	March	April	May	June	July	Aug.	Sept.	Oct.	Nov.	Dec.	Annual
1898	....	....	....	....	....	....	....	....	....	....	70	67	....
1899	65	61	72	82	89	93	93	92	89	78	70	68	93
1900	63	65	67	78	90	90	97	96	91	85	78	66	97
1901	63	55	78	86	86	102	104	94	93	81	67	70	104
1902	50	60	70	89	92	98	100	92	93	80	78	60	100
1903	52	71	77	92	94	90	98	95	89	83	75	52	98
1904	58	60	72	84	91	97	....	95	93	85	63	60	97
1905	62	50	82	83	89	95	98	95	93	89	70	60	98
1906	74	65	62	90	95	....	....	....	....	80	67	66	....
1907	71	50	90	82	84	90	92	94	90	77	64	64	94
1908	60	66	81	87	92	96	100	95	86	86	71	69	100
1909	63	71	73	85	90	96	93	97	86	81	75	58	97
1910	57	69	83	84	88	94	94	93	94	87	66	59	94
1911	59	66	72	78	97	97	105	97	88	78	69	66	105
1912	55	59	71	80	87	91	96	95	95	84	74	67	96
1913	65	72	75	81	91	96	101	93	91	77	75	60	101
1914	69	59	73	80	94	100	99	98	93	85	77	64	100
1915	62	69	59	94	85	91	95	99	94	79	70	63	99
1916	69	64	69	83	90	88	91	99	91	86	73	67	99
1917	57	63	75	85	86	92	96	96	83	80	67	51	96
1918	59	63	76	79	89	95	99	103	84	81	69	66	103
1919	62	64	74	75	90	95	102	90	93	89	72	65	102
1920	49	52	75	80	82	94	93	91	87	84	72	62	94
1921	60	69	85	85	88	97	98	95	95	78	77	62	98
1922	55	72	75	85	85	90	93	89	91	89	72	59	93
1923	62	54	78	82	90	101	99	98	89	82	64	65	101
1924	66	58	74	79	86	97	95	100	97	82	75	69	100
1925	52	68	74	83	97	103	96	93	92	78	69	59	103

AT ROCK HALL (No. 1)—(Shore Station).

Year	Jan.	Feb.	March	April	May	June	July	Aug.	Sept.	Oct.	Nov.	Dec.	Annual
1898	....	....	72	....	88	96	100	92	96	85	65	....	100
1899	54	58	65	....	....	97	94	94	....	....	....	....	97
1900	....	....	....	76	94	....	....	....	....	....	....	....	...

TABLE II.—Continued
AT ROCK HALL (NO. 2).

Year	Jan.	Feb.	March	April	May	June	July	Aug.	Sept.	Oct.	Nov.	Dec.	Annual
1898	....	....	74	77	89	94	99	91	94	85	66	62	99
1899	54	58	66	79	89	95	92	93	90	79	69	66	95
1900	57	61	65	76	91	92	98	101	91	84	73	60	101
1901	60	56	73	78	79	97	102	91	89	79	65	66	102
1902	51	60	....	....	....	....	....	....	....	....	....	....	....
1919	....	....	....	....	....	....	....	....	92	90	71	63	....
1920	48	54	....	....	....	....	92	....	....	....	....	....	....
1921	....	....	....	....	....	....	96	95	95	79	77	65	96
1922	54	65	74	86	87	91	94	91	91	88	71	56	94
1923	62	57	76	82	87	95	95	97	88	81	65	63	97
1924	61	59	72	77	83	96	97	99	95	79	74	68	99
1925	51	67	73	85	92	101	96	91	92	76	69	61	101

TABLE III.*

LOWEST TEMPERATURES AT BETTERTON.

1898	....	....	27	25	42	52	58	....	....	....	....	....	....

AT CHESTERTOWN.

Year	Jan.	Feb.	March	April	May	June	July	Aug.	Sept.	Oct.	Nov.	Dec.	Annual
1856	....	....	....	....	....	....	....	....	....	....	....	....	3
1858	....	....	....	....	....	....	....	....	....	....	....	....	7
1859	....	....	....	....	....	....	....	....	....	....	....	....	0
1861	....	....	....	....	....	....	....	....	....	....	....	....	1
1862	....	....	....	....	....	....	....	....	....	....	....	....	10
1864	....	....	....	....	....	....	....	....	....	....	....	....	5
1893	....	....	....	....	....	....	....	....	....	....	22	12	....
1894	20	9	....	....	41	46	55	51	44	38	....	9	9
1895	11	0	22	31	41	54	55	....	....	30	25	15	0
1896	8	6	16	29	41	50	58	53	41	32	28	9	6
1897	8	11	20	30	45	43	60	59	43	36	26	....	8
1898	....	9	24	25	37	50	54	57	49	31	25	16	9
1899	5	—9	25	28	43	52	54	58	43	32	27	9	—9
1900	....	....	....	....	41	53	57	....	47	34	28	15	....
1901	10	12	14	35	43	....	63	57	44	35	21	11	10
1902	13	8	20	31	40	50	58	54	46	32	29	15	8
1903	11	5	25	27	32	48	54	56	43	35	17	11	5
1904	—2	3	18	28	44	47	54	51	37	30	22	7	—2
1905	—6	0	18	29	42	48	58	54	43	34	20	21	—6
1906	13	8	20	27	36	53	55	62	48	34	30	12	8
1907	8	3	20	23	36	46	56	54	43	31	29	19	3
1908	10	3	25	29	41	51	58	51	46	37	26	13	3
1909	11	15	20	27	38	52	55	53	44	30	29	9	9
1910	8	6	27	36	39	46	53	55	50	31	24	4	4
1911	11	20	9	26	39	53	58	56	44	36	23	22	9
1912	—7	5	18	28	39	47	52	52	43	37	22	16	—7
1913	23	11	16	32	39	45	58	52	43	34	30	22	11

AT COLEMAN.

Year	Jan.	Feb.	March	April	May	June	July	Aug.	Sept.	Oct.	Nov.	Dec.	Annual
1898	....	....	25	25	39	50	55	56	50	33	24	15	....
1899	4	—10	24	28	42	51	54	57	42	33	30	....	—10
1900	....	....	....	....	....	....	....	....	....	....	....	....	....
1901	....	....	....	....	....	....	....	....	....	....	....	....	....
1902	....	9	19	31	42	....	....	....	....	....	28	16	9
1903	10	3	27	26	35	49	57	53	42	35	18	10	3

* Figures in italic denote interpolated data.

TABLE III.—Continued

Year	Jan.	Feb.	March	April	May	June	July	Aug.	Sept.	Oct.	Nov.	Dec.	Annual
1904	—2	1	18	26	43	47	56	52	37	29	22	9	—2
1905	—1	1	16	30	41	50	60	53	41	35	21	19	—1
1906	14	8	20	29	39	55	58	62	51	35	29	12	8
1907	8	5	20	24	35	47	58	57	44	31	27	19	5
1908	....	....	....	28	40	51	59	52	46	38	24	24	....
1909	9	15	21	27	39	54	54	54	43	32	30	8	8
1910	....	....	....	35	43	45	55	56	50	32	25	8	....
1911	13	19	15	22	39	53	59	55	47	36	23	23	13
1912	—6	5	17	....	41	48	56	54	44	40	26	16	—6
1913	23	12	17	34	38	47	59	56	45	35	30	21	12
1914	3	5	15	28	40	52	58	56	45	....	....	....	3
1915	16	17	22	31	....	....	....	....	....	....	....	....	....
1916	....	....	....	....	....	....	....	....	44	38	25	11	....
1917	....	2	21	26	41	52	61	56	44	30	22	—3	—3
1918	2	—6	21	....	42	52	56	57	46	39	26	24	—6
1919	12	20	28	24	45	....	....	59	50	42	29	5	5
1920	7	4	15	30	37	50	56	57	47	39	24	18	4
1921	9	18	25	28	43	49	62	56	56	37	32	11	9
1922	12	4	23	34	40	57	58	55	45	33	29	15	4
1923	17	10	16	13	39	51	56	52	47	37	28	23	10.
1924	6	15	24	28	43	50	59	53	46	35	21	11	6
1925	4	16	10	31	38	52	53	52	40	29	25	12	4

AT GALENA.

Year	Jan.	Feb.	March	April	May	June	July	Aug.	Sept.	Oct.	Nov.	Dec.	Annual
1888	....	....	....	....	....	....	....	....	46	42	28	15	....
1889	25	6	32	41	50	58	68	63	52	36	29	24	6
1890	20	25	11	37	44	62	....	....	....	....	....	....	....

AT MILLINGTON.

Year	Jan.	Feb.	March	April	May	June	July	Aug.	Sept.	Oct.	Nov.	Dec.	Annual
1898	....	....	....	....	....	....	....	....	....	33	25	15	....
1899	5	—7	25	29	42	54	53	55	43	32	25	8	—7
1900	12	5	11	28	37	51	51	56	46	34	26	10	5
1901	10	9	11	32	40	47	63	60	40	32	20	9	9
1902	11	5	20	30	39	49	54	50	41	30	27	12	5
1903	11	5	25	27	34	45	52	51	38	33	16	11	5
1904	—4	2	18	27	40	45	55	48	34	26	21	0	—4
1905	—10	—3	18	29	40	46	52	50	39	34	17	16	—10
1906	10	8	18	27	36	....	....	....	45	30	23	11	8
1907	6	—5	18	23	34	44	54	50	40	28	23	19	—5
1908	9	0	24	27	40	48	53	47	40	33	23	6	0
1909	7	14	16	25	34	47	49	48	39	29	27	4	4
1910	0	5	20	32	35	44	49	54	45	29	20	—3	—3
1911	3	17	3	24	35	50	54	51	46	33	23	22	3
1912	—10	5	14	28	27	42	51	47	39	30	23	8	—10
1913	22	13	14	31	33	42	53	51	41	32	27	20	13
1914	3	1	2	25	37	43	55	56	39	28	18	4	1
1915	16	17	22	28	41	46	56	51	38	33	22	17	16
1916	6	2	11	32	41	45	54	50	39	29	20	0	0
1917	13	1	22	27	37	51	59	49	37	27	16	—4	—4
1918	1	—12	20	28	41	48	49	51	39	29	22	20	—12
1919	10	15	26	23	42	47	50	53	44	36	22	—1	—1
1920	7	5	11	29	32	50	49	53	38	38	18	18	5
1921	9	15	24	27	41	44	58	50	50	31	25	11	9
1922	6	6	21	29	34	49	56	50	37	27	24	12	6
1923	15	7	17	14	35	48	49	47	37	31	20	19	7
1924	6	12	23	25	34	47	50	48	39	25	19	9	6
1925	—1	17	11	24	33	48	49	46	38	27	21	7	—1

TABLE III.—Continued
AT ROCK HALL (No. 1)—(Shore Station).

Year	Jan.	Feb.	March	April	May	June	July	Aug.	Sept.	Oct.	Nov.	Dec.	Annual
1898	...	...	24	29	40	52	49	57	49	31	26	17	...
1899	1	—6	26	...	...	56	53	54	...	...	...	...	—6
1900	...	...	...	28	39	...	...	...	...	...	...	...	...

AT ROCK HALL (No. 2).

Year	Jan.	Feb.	March	April	May	June	July	Aug.	Sept.	Oct.	Nov.	Dec.	Annual
1898	...	...	22	25	39	51	49	54	43	32	24	11	...
1899	1	—6	25	25	41	50	51	56	41	28	23	11	—6
1900	10	5	13	28	39	47	55	55	43	32	24	14	5
1901	11	11	12	36	41	48	64	54	40	29	17	8	8
1902	10	2	...	...	...	...	...	...	...	...	...	...	2
1919	...	...	...	...	...	...	...	...	44	37	25	2	...
1920	10	7	...	...	...	...	50	...	...	...	...	...	7
1921	...	...	...	...	...	...	60	53	50	34	27	12	...
1922	5	9	23	30	35	56	57	53	38	30	26	13	5
1923	16	11	17	14	37	51	53	47	39	34	23	21	11
1924	8	13	24	28	35	48	54	51	41	28	16	10	8
1925	3	18	11	29	33	49	50	47	40	28	22	10	3

TABLE IV.
MEAN TEMPERATURE RANGE AT CHESTERTOWN.

	Jan.	Feb.	March	April	May	June	July	Aug.	Sept.	Oct.	Nov.	Dec.	Annual
Highest	43.0	43.2	49.2	56.8	69.2	75.3	79.0	77.0	72.2	60.8	50.6	41.8	57.5
Lowest	21.4	23.8	37.4	47.0	58.4	65.7	72.0	71.5	65.0	50.2	39.6	28.6	51.0
Range	21.6	19.4	11.8	9.8	10.8	9.6	7.0	5.5	7.2	10.6	11.0	13.2	6.5
Extreme Max.	70	71	82	87	94	96	100	95	94	84	77	74	100
Extreme Min.	—7	—9	9	23	32	43	52	51	37	30	17	4	—9
Range	77	80	73	64	62	53	48	44	57	54	60	70	109

AT COLEMAN.

	Jan.	Feb.	March	April	May	June	July	Aug.	Sept.	Oct.	Nov.	Dec.	Annual
Highest	42.6	43.6	54.2	58.5	68.2	77.8	79.3	77.6	73.9	63.1	51.8	43.9	57.8
Lowest	26.6	25.2	36.5	48.2	58.8	66.2	75.0	72.0	65.0	52.6	42.2	28.3	52.3
Range	16.0	18.4	17.7	10.3	9.4	11.6	4.3	5.6	8.9	10.5	9.6	15.6	5.5
Extreme Max.	72	70	87	90	96	100	104	106	95	92	76	69	106
Extreme Min.	—6	—10	10	13	35	45	54	52	37	29	18	—3	—10
Range	78	80	77	77	61	55	50	54	58	63	58	72	116

AT MILLINGTON.

	Jan.	Feb.	March	April	May	June	July	Aug.	Sept.	Oct.	Nov.	Dec.	Annual
Highest	42.5	42.8	55.2	58.8	68.7	77.0	80.4	77.3	72.6	62.4	51.4	43.7	57.0
Lowest	22.6	25.6	36.4	47.8	57.4	65.2	73.6	71.0	62.6	52.0	40.6	28.1	52.2
Range	19.9	17.2	18.8	11.0	11.3	11.8	6.8	6.3	10.0	10.4	10.8	15.6	4.8
Extreme Max.	74	72	90	94	97	103	105	103	97	89	78	70	105
Extreme Min.	—10	—12	2	14	32	42	49	46	34	25	16	—4	—12
Range	84	84	88	80	65	61	56	57	63	64	62	74	117

AT ROCK HALL (No. 2).

	Jan.	Feb.	March	April	May	June	July	Aug.	Sept.	Oct.	Nov.	Dec.	Annual
Highest	37.5	41.0	55.0	59.0	65.7	77.2	79.5	79.0	72.6	62.7	49.4	44.2	57.9
Lowest	27.8	26.6	39.4	50.1	59.6	70.0	74.6	72.4	64.4	52.6	41.0	31.8	53.7
Range	9.7	14.4	15.6	8.9	6.1	7.2	4.9	6.6	8.2	10.1	8.4	12.4	4.2
Extreme Max.	62	67	76	86	92	101	102	101	95	90	77	68	102
Extreme Min.	1	—6	11	14	33	47	49	47	38	28	16	2	—6
Range	61	73	65	72	59	54	53	54	57	62	61	66	108

TABLE V.

KILLING FROSTS AT CHESTERTOWN.

	Last in Spring.	First in Autumn.
1895 :	April 12	October 22
1896 :	April 9	October 25
1897 :	April 21	November 14
1898 :	April 28	October 28
1899 :	April 11	October 22
1900 :	April 15	October 20
1901 :	March 30	November 4
1902 :	April 4	October 30
1903 :	May 2	November 7
1904 :	April 20	October 31
1905 :	April 19	October 22
1906 :	April 3	October 12
1907 :	April 21	October 31
1908 :	April 17	October 13
1909 :	April 12	October 20
1910 :	March 19	October 30
1911 :	April 12	November 3
1912 :	April 9	November 3
1913 :	May 12	October 22
Average :	April 14	October 27

AT COLEMAN.

	Last in Spring.	First in Autumn.
1898 :	April 7	October 28
1899 :	April 11	October 22
1900 :	April 11	October 20
1901 :	March 29	October 26
1902 :	April 4	October 30
1903 :	April 6	October 29
1904 :	April 20	October 28
1905 :	April 19	October 22
1906 :	April 3	October 12
1907 :	April 15	October 31
1908 :	April 17	November 5
1909 :	April 12	October 20
1910 :	April 14	October 30
1911 :	April 10	November 3
1912 :	April 9	November 4
1913 :	May 12	October 22
1914 :	April 14	October 28
1915 :	April 4	October 11
1916 :	April 11	October 11
1917 :	April 14	October 13
1918 :	April 6	November 7
1919 :	April 2	November 10
1920 :	April 11	November 13
1921 :	April 2	November 11
1922 :	April 24	October 21
1923 :	April 10	November 2
1924 :	April 3	October 23
1925 :	April 7	October 28
Average :	April 11	October 27

TABLE V.—Continued

AT MILLINGTON.

	Last in Spring.		First in Autumn.	
1898 :	April	28	October	28
1899 :	April	10	October	22
1900 :	April	11	November	6
1901 :	April	4	October	26
1902 :	April	4	October	30
1903 :	April	5	November	7
1904 :	April	20	October	8
1905 :	April	19	November	2
1906 :	April	3	October	12
1907 :	April	20	October	15
1908 :	April	20	October	13
1909 :	April	12	October	17
1910 :	April	14	October	30
1911 :	April	18	October	29
1912 :	April	9	October	17
1913 :	May	12	October	22
1914 :	April	14	October	28
1915 :	April	5	November	6
1916 :	April	11	October	11
1917 :	April	15	October	7
1918 :	April	7	October	23
1919 :	April	25	November	7
1920 :	May	6	November	13
1921 :	April	12	October	14
1922 :	April	29	October	19
1923 :	April	10	October	7
1924 :	April	5	October	22
1925 :	April	22	October	11
Average :	April	15	October	23

AT ROCK HALL.

	Last in Spring.		First in Autumn.	
1898 :	April	9	October	28
1899 :	April	11	October	22
1900 :	April	11	October	20
1901 :	March	19	October	7
1902 :				
1919 :	April	2	November	9
1920 :	April	11	November	13
1921 :	April	12	November	11
1922 :	April	29	October	21
1923 :	April	10	November	2
1924 :	April	3	October	23
1925 :	April	21	October	28
Average :	April	9	October	28

TABLE VI.*

MONTHLY AND ANNUAL PRECIPITATION AT BETTERTON.

Year	Jan.	Feb.	March	April	May	June	July	Aug.	Sept.	Oct.	Nov.	Dec.	Annual
1898	....	....	2.33	1.90	2.62	0.40	4.37	....	....	....	....	....	.....

AT CHESTERTOWN.

Year	Jan.	Feb.	March	April	May	June	July	Aug.	Sept.	Oct.	Nov.	Dec.	Annual
1855	....	....	....	....	4.10	3.03	9.44	....	4.89	1.65	4.14	....	.....
1856	2.40	0.87	....	....	....	....	....	....	....	....	....	....	.....
1857	....	....	....	....	....	....	4.28	....	....	....	....	....	.....
1858	1.94	1.72	0.76	3.98	6.86	2.36	1.69	....	2.64	1.54	4.59	1.07	.....
1859	....	....	5.99	....	....	3.12	2.20	....	....	3.13	....	....	.....
1860	....	....	....	....	....	....	....	....	....	....	....	....	.....
1861	4.55	2.30	3.38	3.80	5.73	4.74	2.85	4.58	4.10	3.01	4.03	1.65	44.32
1862	4.24	4.24	3.34	6.18	2.04	9.16	3.22	....	....	2.52	3.85	1.25	.....
1863	....	....	....	....	4.09	1.91	5.56	....	....	2.08	1.66	4.34	.....
1864	2.21	0.37	2.51	4.82	3.39	1.37	1.05	....	....	....	....	....	.....
1893	....	....	....	....	....	....	....	....	....	2.	2.60	....	.....
1894	2.34	.80	.20	4.2	6.67	.40	1.97	3.54	.83	.28	.9	....	.81
1895	4.	.	.20	6.0	.4	.86	3.45	2.	.	.	.4	.	.14
1896	2.	.	.91	1:1	.0	.73	4.87	2.	.	.	.8	.	.02
1897	2.	.	.7	3.5	.7	.35	8.43	4.	.	.	.00	.	.86
1898	3.	.	.9	2.7	.7	.17	3.05	7.	.	.	.98	.	.39
1899	3.	.	.9	1.2	.8	.83	4.53	4.	.	.	.2	.	.71
1900	2.	.	.2	2.32	.6	.90	2.37	2.	.	.	.5	.	.48
1901	2.	.	.7	5.70	.9	.38	8.48	6.	.	.	.4	.	.32
1902	3.	.	.0	3.9	.2	.48	4.07	1.	.	.	.8	.	.15
1903	3.	.	.1	3.2	.8	.86	5.55	4.	.	.	.4	.	.87
1904	3.	.	.6	1.9	.0	.	5.38	2.	.	.	.5	.	.27
1905	4.	.	.1	3.3	.1	.	10.00	4.	.	.	.5	.	.02
1906	2.	.	.1	2.6	.9	.	5.14	8.	.	.	.88	.	.91
1907	2.	.	.7	4.1	.7	.	4.67	3.	.	.	.83	.	.14
1908	3.	.	.1	2.4	.7	.	2.07	6.	.	.	.2	.	.34
1909	3.	.	.3	2.1	.2	.	2.05	1.00	.	.	.1	.	.70
1910	4.	.	.0	3.7	.2	.	3.56	4.21	.	.	.7	.	.98
1911	3.	.	.0	3.5	T	.	3.71	10.07	.	.	.3	.	.23
1912	3.	.	.4	2.3	4.24	.	4.92	1.87	.	.	.5	.	.46
1913	4.	.	.4	6.8	3.88	.	1.47	4.06	.	.	.1	.	.52
Av.	3.28	3.09	3.61	3.59	3.74	3.96	4.06	4.59	3.56	3.10	2.85	3.42	42.85

AT COLEMAN.

Year	Jan.	Feb.	March	April	May	June	July	Aug.	Sept.	Oct.	Nov.	Dec.	Annual
1898	3.40	2.00	2.69	2.47	4.34	0.85	5.15	6.34	2.27	5.36	4.49	5.28	44.64
1899	.81	.85	.	.	2.28	.05	.	5.78	.45	.	.28	1.70	47.13
1900	.	.6	.	.	3.04	.6	.	4.00	.1	.	.83	3.22	40.99
1901	.	.7	.	.	3.53	.4	.	2.70	.0	.	.72	5.81	39.26
1902	.	.2	.	.	1.5	.20	.	1.50	.00	.	.89	7.42	51.92
1903	.00	.4	.	.	1.0	.09	.	6.00	.35	.	.79	3.76	45.08
1904	.64	.04	.	.	3.6	.7	.	2.77	.14	.	.01	3.34	38.83
1905	.	.46	.	.	2.5	.3	.	5.0	.06	.	.39	4.07	42.43
1906	.	.49	.	.	4.8	.9	.	6.5	.68	.	.25	4.19	50.43
1907	.	.5	.	.	4.2	.4	.	5.3	.	.	.45	5.50	56.08
1908	.	.00	.	.	5.5	.06	.	5.4	.	.	.17	3.66	39.26
1909	.	.75	.	.	3.2	.12	.	0.7	.	.28	.74	5.47	34.86
1910	.	.60	.	.	3.7	.63	.	3.4	.	.12	.74	2.35	36.30
1911	.	.82	.	.	2.1	.	.	12.04	.	.93	.51	4.16	48.44
1912	.	.92	.	.	3.6	.	.	2.39	.	.	.70	4.88	43.48

* Figures in italic denote interpolated data.

TABLE VI.—Continued

Year	Jan.	Feb.	March	April	May	June	July	Aug.	Sept.	Oct.	Nov.	Dec.	Annual
1913	3.19	1.61	5.34	7.11	3.53	1.33	2.16	4.15	2.14	3.88	1.82	2.23	38.49
1914		.48	1.7	.05			2.4	4.72	.9	1.0	2.35	4.20	30.46
1915	.	.4	1.0	.6	.		2.5	9.00	0	3.0	1.40	3.00	42.08
1916	.	.20	3.8	.00	.		6.80	2.00	0		09	5.00	39.02
1917	.	.75	5.3	.28	.		5.73	1.82	.0	.	45	1.33	39.79
1918	.	.2	3.9	.78	.06	.	3.4	4.23	.1	.	9	4.	40.74
1919	.	.7	4.8	.1	.46	.	11.2	6.94	.9	.	8	3.	57.44
1920	.	.2	3.2	.06		.	5.5	11.	.9	.	3	3.	49.13
1921	.	.7	1.7	.07	.	.	5.3	3.	.2	.	05	1.	38.11
1922	.	.4	4.2	.8	.	.	8.9	1.	.9	.	46	3.	39.95
1923	.	.7	3.8	.	.	.	5.1	2.	.0	.	7	3.09	38.39
1924	.	.1	3.5	.	.	.	0.2	4.	.8	.	0	2.47	42.89
1925	.	.5	2.0	.	.	7	6.4	1.	.3	.	4	1.32	30.31
Av.	3.45	3.07	3.64	3.53	3.38	3.65	4.62	4.51	3.41	2.81	2.52	3.76	42.35

AT GALENA.

Year	Jan.	Feb.	March	April	May	June	July	Aug.	Sept.	Oct.	Nov.	Dec.	Annual
1888										3.97	3.41	8.04	
1889	4.49	2.48	4.45	5.97	6.08	5.84	8.46	2.09	4.63	4.75	10.17	0.55	59.96
1890	1.63	3.69	4.75	3.34	3.90	1.60							

AT MILLINGTON.

Year	Jan.	Feb.	March	April	May	June	July	Aug.	Sept.	Oct.	Nov.	Dec.	Annual
1899	3.30	5.54	4.78	1.77	2.68	2.56	4.24	5.15	3.58	1.65	2.17	1.59	39.01
1900	3.73	6.89	3.41	2.	2.02	.4	4.9	.1	.82	2.			
1901	3.12	0.64	2.68	5.	.83	.4	8.0	.5	.38	1.			
1902	4.25	6.85	3.30	3.	2.28	.0	3.7	.7	.68	4.			
1903	3.08	4.71	6.35	3.	1.41	.3	4.3	.	.61	5.			
1904	2.89	2.80	3.58	2.	2.25	.8	4.0	.	.47	2.			
1905	4.14	3.66	3.78	3.	3.79	.3	7.8	.	.88	1.			
1906	2.66	3.88	6.73	2.	2.19	.00	7.00	.	.75	4.			
1907	2.33	2.50	2.58	3.	6.82	.79	3.05	.	.28	2.			
1908	3.14	3.77	2.21	2.	7.00	.3	5.6	.	.38	2.			
1909	3.17	3.62	4.26	2.	3.02	.0	2.6	.	.71	1.			
1910	4.80	1.24	1.89	4.	2.49	.5	3.5	.	.88	4.			
1911	4.41	2.49	3.42	4.	0.39	.4	2.3	.	.15	3.			
1912	3.78	1.99	8.16	2.	4.18	.3	2.7	.7	.43	3.			
1913	3.57	1.53	4.44	7.	2.66	.3	1.5	.5	.01	5.			
1914	3.91	2.95	3.44	3.	2.33	.3	4.2	.4	.49	1.			
1915	5.26	4.80	1.58	3.	3.23	.2	2.4	.4	.81	4.			
1916	1.87	3.94	4.11	3.	3.08	.8	5.6	.6	.19	1.			
1917	3.03	1.98	5.48	2.	2.92	.7	6.4	.3	.40	7.			
1918	4.54	1.04	5.38	5.	4.14	.7	5.1	.4	.37	1.			
1919	3.64	2.44	4.22	4.	5.88	.3	10.8	.3	.05	2.			
1920	2.57	3.58	3.20	4.94	2.61	.9	2.8	.2	.12	1.			
1921	2.25	2.94	2.49	3.95	3.73	.1	3.	.9	.24	0.			
1922	3.55	3.83	4.45	1.8	2.37	.3	7.	.4	.55	1.			
1923	4.22	2.53	4.87	4.	1.66	.1	2.	.2	.89	3.			
1924	4.55	4.48	5.82	5.	6.40	.0	1.	.0	.97	0.	.23		
1925	5.19	1.75	2.37	2.	3.25	.4	8.	.8	.66	4.	.05		
Av.	3.59	3.25	4.04	3.62	3.23	3.79	4.70	4.34	3.58	2.91	2.57	3.77	43.39

AT ROCK HALL (No. 1)—(Shore Station).

Year	Jan.	Feb.	March	April	May	June	July	Aug.	Sept.	Oct.	Nov.	Dec.	Annual
1898			2.74	2.27	5.47	0.91	4.04	6.97	1.68	4.49	4.29	2.56	
1899	3.77	5.57	5.07			4.85	3.43	5.44					
1900				2.02	2.44								

TABLE VI.—Continued

AT ROCK HALL (No. 2).

Year	Jan.	Feb.	March	April	May	June	July	Aug.	Sept.	Oct.	Nov.	Dec.	Annual
1898	....	....	2.51	1.77	5.28	1.49	3.43	7.69	2.32	3.70	4.52	3.36	.....
1899	3.77	5.27	4.99	1.03	2.57	3.84	3.52	5.48	3.37	1.77	2.59	1.03	39.23
1900	2.63	4.91	2.72	2.10	2.82	4.66	1.81	4.26	8.23	1.49	2.01	2.22	39.86
1901	2.41	0.31	1.75	5.18	3.22	1.43	8.31	4.10	4.35	1.08	3.30	6.71	42.15
1902	3.38	4.09	....	....	....	....	....	....	....	....	....	....	.....
1919	....	....	....	....	....	....	....	2.45	.45	.5	3.55	3.30	.....
1920	2 33	3.58	3.20	4 90	2.65	6.0	5.00	9 4	..	.0	3.00	3 20	46.46
1921	2 50	.90	2.40	3 20	3.70	2.0	4.50	2	.	.5	4.10	2 26	34.12
1922	4 57	20	6.01	1 95	2.18	5.6	7.0	2	.	.0	0.50	3 67	41.63
1923	4 1	.8	4.22	4 8	2.0	3.1	5.3	3	.	.1	2.31	2 80	42.71
1924	3 6	.8	5.15	5 7	5.6	4.2	1.3	4	.	.6	2.02	2 82	46.76
1925	4 7	.3	2.50	2 6	1.8	1.8	7.3	1	.	.6	3.35	1 73	34.54
Av.	3.41	3.24	3.54	3.33	3.20	3.42	4.76	4.50	3.89	2.03	2.84	3.01	41.17

TABLE VII.

NUMBER OF DAYS WITH .01 INCH OR MORE OF PRECIPITATION AT CHESTERTOWN.

(Rainfall and Melted Snow.)

Year	Jan.	Feb.	March	April	May	June	July	Aug.	Sept.	Oct.	Nov.	Dec.	Annual
1893	..	..	..	..	11	4	7	6	..	..	5	6	....
1894	7	7	....	..	8	8	8	6	7	6	....	5	....
1895	8	3	....	8	8	8	6	..	..	3	5	5	....
1896	2	9	9	3	10	9	10	4	6	6	5	5	78
1897	6	11	12	9	11	11	12	9	4	13	11	..	....
1898	..	3	8	10	10	5	8	8	3	9	10	8	....
1899	7	12	11	4	10	5	11	12	9	6	....	7	....
1900	..	..	..	..	7	6	7	..	8	8	7	8	....
1901	7	4	8	15	13	....	14	10	8	5	4	8	....
1902	8	8	9	9	7	11	10	6	12	6	10	12	108
1903	9	12	10	11	4	16	15	14	4	7	7	9	118
1904	11	10	14	8	8	12	13	10	7	4	6	10	113
1905	10	8	11	10	11	12	14	7	6	7	6	8	110
1906	7	5	14	5	9	12	15	14	6	12	7	....	....
1907	12	8	9	13	14	13	10	5	12	7	12	8	123
1908	8	11	11	10	13	2	9	8	3	7	3	11	96
1909	8	14	10	8	10	12	4	4	4	6	9	9	98
1910	11	9	7	6	8	16	6	11	4	5	5	4	92
1911	10	7	12	10	0	9	9	13	6	11	10	9	106
1912	11	7	11	11	8	7	8	6	13	2	4	12	100
1913	15	6	12	11	5	4	5	8	6	8	6	4	90
Av.	9	8	10	9	9	9	10	9	7	7	7	8	101

AT COLEMAN.

Year	Jan.	Feb.	March	April	May	June	July	Aug.	Sept.	Oct.	Nov.	Dec.	Annual
1898	....	5	12	9	12	7	7	9	3	9	12	7	....
1899	8	9	12	3	11	5	7	9	7	5	5	7	88
1900	7	9	8	5	6	6	8	....	4	3	5	5	....
1901	6	2	9	12	13	4	12	7	7	3	5	9	89
1902	5	8	10	7	6	....	....	....	....	....	7	9	....
1903	9	10	10	9	6	9	9	12	4	6	5	7	96
1904	10	7	9	9	9	11	9	6	5	2	5	8	90
1905	6	6	8	9	8	11	14	8	5	4	2	7	88
1906	10	6	11	4	7	10	11	11	3	9	4	9	95
1907	11	5	9	7	14	11	9	11	9	7	9	8	110
1908	....	....	....	6	9	1	6	5	2	5	3	6	....
1909	6	11	9	6	5	11	3	3	6	4	7	5	76
1910	....	....	....	....	14	12	6	8	3	5	4	7	....
1911	12	6	9	8	4	11	7	15	5	13	10	6	106
1912	9	5	10	9	8	7	10	8	8	4	4	8	90
1913	14	7	9	10	5	5	6	9	6	7	5	4	87

TABLE VII.—Continued

Year	Jan.	Feb.	March	April	May	June	July	Aug.	Sept.	Oct.	Nov.	Dec.	Annual
1914	5	5	8	7	4	8	8	10	3	....	....	....	....
1915	11	8	3	6	11	....	....	....	....	....	....	....	....
1916	....	....	....	....	....	....	....	....	....	4	10	10	....
1917	10	6	12	5	9	9	11	7	5	8	4	7	93
1918	9	6	10	8	8	6	5	9	8	5	5	10	89
1919	7	9.	10	7	13	7	13	12	2	9	11	11	111
1920	8	9	8	9	7	12	10	15	4	1	8	7	98
1921	7	8	11	9	12	6	9	8	7	4	14	7	102
1922	8	14	14	9	9	16	14	6	5	6	2	11	114
1923	9	11	10	9	4	8	13	10	8	4	7	9	102
1924	6	8	8	11	14	16	4	8	9	1	6	7	98
1925	13	7	8	12	8	4	10	8	6	15	9	8	108
Av.	9	7	9	8	9	9	9	9	5	6	6	8	94

AT MILLINGTON.

Year	Jan.	Feb.	March	April	May	June	July	Aug.	Sept.	Oct.	Nov.	Dec.	Annual
1898	....	....	....	....	....	....	....	....	....	....	....	6	....
1899	7	9	11	3	10	6	8	10	7	2	6	6	85
1900	6	8	11	5	6	4	8	8	6	7	5	4	78
1901	6	3	5	10	11	4	10	8	7	3	5	9	81
1902	6	8	9	8	8	9	....	5	12	6	9	13	....
1903	7	7	7	7	4	10	9	10	5	6	4	5	81
1904	11	8	10	8	8	10	8	10	3	3	....	9	....
1905	6	7	9	7	5	7	10	7	7	4	5	9	83
1906	11	6	14	5	7	....	....	....	10	12	7	11	....
1907	14	5	8	11	15	13	9	8	10	8	13	10	124
1908	10	10	12	12	15	3	10	9	4	10	6	10	111
1909	9	12	12	9	10	17	4	4	8	6	10	9	110
1910	13	11	9	12	12	16	11	13	6	9	9	8	129
1911	13	10	13	18	4	12	6	15	5	11	12	9	123
1912	11	6	16	14	9	11	8	5	12	3	6	11	112
1913	14	8	11	11	11	7	6	10	8	13	8	9	116
1914	6	7	12	11	7	11	12	10	5	6	5	15	107
1915	14	9	4	7	14	9	14	18	6	11	5	7	118
1916	14	9	14	12	12	11	10	6	9	5	11	10	123
1917	13	9	14	7	11	8	14	13	6	10	4	8	117
1918	13	5	10	10	9	8	8	10	9	6	5	11	104
1919	10	8	10	7	13	6	13	15	4	12	11	10	119
1920	10	10	10	9	8	13	9	20	4	2	10	8	113
1921	8	10	10	13	14	7	14	8	4	4	17	12	125
1922	9	13	13	10	11	15	14	13	5	8	8	13	132
1923	11	11	12	10	5	11	12	12	8	7	9	12	120
1924	6	8	10	13	17	18	5	8	14	2	7	11	119
1925	14	9	9	12	10	6	12	6	7	14	8	8	115
Av.	10	8	11	9	10	10	9	10	7	7	8	9	108

ROCK HALL (No. 1)—(Shore Station).

Year	Jan.	Feb.	March	April	May	June	July	Aug.	Sept.	Oct.	Nov.	Dec.	Annual
1898	....	....	7	9	16	7	10	9	4	10	12	10	....
1899	12	14	12	....	....	8	10	10	....	....	....	....	....
1900	....	....	....	8	6	....	....	....	....	....	....	....	....

AT ROCK HALL (No. 2).

Year	Jan.	Feb.	March	April	May	June	July	Aug.	Sept.	Oct.	Nov.	Dec.	Annual
1898	....	....	12	11	18	5	8	10	4	12	13	10	....
1899	12	14	11	5	11	8	11	9	8	4	6	6	105
1900	8	9	10	8	8	6	7	9	10	8	9	7	99
1901	7	3	10	13	13	6	15	11	14	6	5	10	113
1902	9	9	....	....	....	....	....	....	....	....	....	....	....
1919	....	....	....	....	....	....	....	....	4	5	9	10	....

TABLE VII.—Continued

Year	Jan.	Feb.	March	April	May	June	July	Aug.	Sept.	Oct.	Nov.	Dec.	Annual
1920	9	11	9	9	7	12	8	15	4	2	9	7	102
1921	6	8	10	11	12	5	8	8	8	4	13	9	102
1922	9	16	11	9	15	13	12	4	6	4	11	9	119
1923	9	11	12	11	4	10	13	11	9	5	9	9	113
1924	7	8	11	12	15	14	9	7	12	1	5	9	110
1925	12	7	9	11	8	7	12	7	7	15	10	9	114
Av.	9	10	10	10	10	9	10	10	8	6	8	9	109

TABLE VIII.*

MONTHLY AND ANNUAL SNOWFALL AT CHESTERTOWN.

Year	Jan.	Feb.	March	April	May	June	July	Aug.	Sept.	Oct.	Nov.	Dec.	Annual
1893	....	....	....	....	....	....	....	....	....	....	T	3.0	....
1894	2.0	6.0	....	....	....	....	....	....	....	0	....	1.5	....
1895	4.0	13.5	....	0	....	....	....	....	....	0	0	0.2	....
1896	1.0	T	2.0	0	....	....	....	....	....	0	1.0	6.0	10.0
1897	7.5	4.5	T	0	....	....	....	....	....	0	T	....	....
1898	....	0	T	2.0	....	....	....	....	....	0	7.0	3.0	....
1899	3.5	34.2	4.0	0	....	....	....	....	....	0	0	1.5	43.2
1900	....	....	....	....	....	....	....	....	....	0	0	1.7	....
1901	12.1	0.5	T	0	....	....	....	....	....	0	1.0	T	13.6
1902	8.7	6.5	0.5	0	....	....	....	....	....	0	0	3.5	19.2
1903	0.8	0.8	0	0	....	....	....	....	....	T	0.2	3.8	5.6
1904	15.0	4.8	3.0	0	....	....	....	....	....	0	3.0	13.5	39.3
1905	26.7	3.6	T	T	....	....	....	....	....	0		3.0	33.3
1906	1.5	3.5	6.5	T	....	....	....	....	....	0	1.0	T	12.5
1907	4.8	13.5	6.7	T	....	....	....	....	....	0	T	5.0	30.0
1908	8.0	8.7	2.0	0	....	....	....	....	....	0	0.5	8.0	27.2
1909	3.2	1.5	13.0	T	....	....	....	....	....	0	2.0	16.0	35.7
1910	6.5	2.0	3.5	0	....	....	....	....	....	0	T	12.0	24.0
1911	5.0	4.0	9.0	T	....	....	....	....	....	0	T	T	18.0
1912	21.0	1.7	8.3	0	....	....	....	....	....	0	0.5	8.0	39.5
1913	T	0.5	T	T	....	....	....	....	....	0	0	T	0.5
Av.	7.3	5.8	3.1	0.1	....	....	....	....	....	T	0.8	4.5	21.6

AT COLEMAN.

Year	Jan.	Feb.	March	April	May	June	July	Aug.	Sept.	Oct.	Nov.	Dec.	Annual
1898	....	....	2.0	2.0	....	....	....	....	....	0	8.2	1.0	....
1899	5.0	31.5	3.0	0	....	....	....	....	....	0	0	1.5	41.0
1900	1.5	11.0	8.0	0	....	....	....	....	....	0	0	2.0	22.5
1901	6.2	2.5	T	0	....	....	....	....	....	0	1.0	0	9.7
1902	9.0	4.0	0.5	0	....	....	....	....	....	....	0	4.5	....
1903	2.0	2.0	0	0	....	....	....	....	....	0	T	5.5	9.5
1904	15.0	5.2	3.0	T	....	....	....	....	....	0	2.0	15.0	40.2
1905	14.0	6.0	T	T	....	....	....	....	....	0	T	3.0	23.0
1906	2.0	5.0	9.0	T	....	....	....	....	....	T	1.0	T	17.0
1907	8.0	12.0	7.5	T	....	....	....	....	....	0	0	4.0	31.5
1908	....	....	....	0	....	....	....	....	....	0	1.0	7.0	....
1909	4.2	1.5	10.0	0	....	....	....	....	....	0	2.0	16.0	33.7
1910	....	....	....	0	....	....	....	....	....	0	....	10.0	....
1911	6.5	8.0	8.5	1.0	....	....	....	....	....	0	T	1.0	25.0
1912	13.5	1.5	9.0	0	....	....	....	....	....	0	1.0	6.0	31.0
1913	1.0	1.0	0	0	....	....	....	....	....	0	0	0	2.0
1914	0	10.0	19.0	0	....	....	....	....	....	....	....	....	....
1915	1.0	0.5	1.5	5.0	....	....	....	....	....	....	....	....	....
1916	....	....	....	....	....	....	....	....	....	0	T	11.0	....
1917	3.1	....	....	5.7	....	....	....	....	....	0	2.0	....	....
1918	20.0	1.0	T	0	....	....	....	....	....	0	0	T	21.0
1919	T	3.3	T	T	....	....	....	....	....	0	0	6.0	9.3
1920	2.0	8.0	5.5	T	....	....	....	....	....	0	0	0.5	16.0
1921	1.0	6.2	0	2.5	....	....	....	....	....	0	T	6.5	16.2

* Figures in italic denote interpolated data.

TABLE VIII.—Continued

Year	Jan.	Feb.	March	April	May	June	July	Aug.	Sept.	Oct.	Nov.	Dec.	Annual
1922	25.5	4.5	0.2	0	....	....	....	....	....	0	T	1.5	31.7
1923	4.5	5.0	5.0	T	....	....	....	....	....	0	0	2.0	16.5
1924	T	7.5	4.0	6.0	....	....	....	....	....	0	0.5	T	18.0
1925	13.0	0	T	0	....	....	....	....	....	3.0	0.5	T	16.5
Av.	6.6	6.0	4.0	0.8	....	....	....	....	....	0.1	0.8	4.2	22.5

AT MILLINGTON.

Year	Jan.	Feb.	March	April	May	June	July	Aug.	Sept.	Oct.	Nov.	Dec.	Annual
1898	....	....	....	....	....	....	....	....	....	....	....	3.0	....
1899	....	24.0	6.0	0	....	....	....	....	....	0	T	1.0	....
1900	T	15.0	5.5	T	....	....	....	....	....	0	0	3.7	24.2
1901	11.5	....	T	0	....	....	....	....	....	0	0.5	T	....
1902	8.0	7.0	....	0	....	....	....	....	....	0	0	2.8	....
1903	2.0	....	0	0	....	....	....	....	....	0	T	6.0	....
1904	10.0	2.5	0.5	T	....	....	....	....	....	0	....	11.5	....
1905	21.0	....	T	T	....	....	....	....	....	0	0	2.0	....
1906	1.0	....	....	0	....	....	....	....	....	0	1.0	T	....
1907	4.5	11.8	5.7	0.3	....	....	....	....	....	0	0	2.4	24.7
1908	7.0	8.6	1.5	T	....	....	....	....	....	0	T	10.0	27.1
1909	6.5	2.0	9.0	T	....	....	....	....	....	0	3.0	25.6	46.1
1910	3.5	3.0	1.8	0	....	....	....	....	....	0	1.5	13.5	23.3
1911	7.0	8.5	7.5	T	....	....	....	....	....	0	0	T	23.0
1912	18.5	2.4	7.8	0	....	....	....	....	....	0	0.5	9.0	38.2
1913	T	2.0	T	0	....	....	....	....	....	0	0	T	2.0
1914	T	5.0	20.5	0	....	....	....	....	....	0	0	1.0	26.5
1915	2.0	3.0	0.4	15.0	....	....	....	....	....	0	T	5.5	25.9
1916	3.7	4.5	7.3	T	....	....	....	....	....	0	T	15.0	30.5
1917	3.5	4.0	0.5	4.5	....	....	....	....	....	0	2.0	6.0	20.5
1918	25.0	2.0	0	0	....	....	....	....	....	0	0	T	27.0
1919	1.0	3.0	T	T	....	....	....	....	....	0	0	6.6	10.6
1920	1.4	9.0	5.4	T	....	....	....	....	....	0	0	0.5	16.3
1921	T	6.2	0	4.0	....	....	....	....	....	0	0	6.3	16.5
1922	21.5	7.0	T	0	....	....	....	....	....	0	T	2.0	30.5
1923	4.0	4.5	5.5	T	....	....	....	....	....	0	0	1.0	15.0
1924	T	9.5	3.0	4.5	....	....	....	....	....	0	0.5	T	17.5
1925	11.3	0	T	0	....	....	....	....	....	3.0	0.5	T	14.8
Av.	6.7	6.3	3.5	1.0	....	....	....	....	....	0.1	0.4	4.8	22.8

AT ROCK HALL (No. 1).

Year	Jan.	Feb.	March	April	May	June	July	Aug.	Sept.	Oct.	Nov.	Dec.	Annual
1898	....	....	1.0	2.0	....	....	....	....	....	....	3.5	0.7	....
1899	3.0	32.5	1.6	....	....	....	....	....	....	....	....	....	....
1900	....	....	....	T	0	....	....	....	....	....	....	....	....

AT ROCK HALL (No. 2).

Year	Jan.	Feb.	March	April	May	June	July	Aug.	Sept.	Oct.	Nov.	Dec.	Annual
1898	....	....	1.0	2.0	....	....	....	....	....	0	3.0	1.	....
1899	4.5	30.5	2.0	0	....	....	....	....	....	0	0	0.0	37.7
1900	2.7	10.0	5.0	0	....	....	....	....	....	0	0	2.0	19.7
1901	4.5	....	T	0	....	....	....	....	....	....	0.5	T	....
1902	8.0	2.0	....	....	....	....	....	....	....	....	....	....	....
1919	....	....	....	....	....	....	....	....	....	0	0	.5	....
1920	2.0	8.0	4.5	T	....	....	....	....	....	0	0	5.5	15.0
1921	T	6.0	0	2.0	....	....	....	....	....	0	0	8.0	16.0
1922	25.5	5.4	T	T	....	....	....	....	....	0	T	1.5	32.4
1923	3.5	5.5	5.0	T	....	....	....	....	....	0	0	2.0	16.0
1924	T	8.5	4.5	6.0	....	....	....	....	....	0	T	T	19.0
1925	16.0	T	T	0	....	....	....	....	....	2.5	0.5	T	19.0
Av.	6.7	8.4	2.2	1.0	....	....	....	....	....	0.2	0.4	1.9	20.8

THE HYDROGRAPHY OF KENT COUNTY

BY

N. C. GROVER

This county lying between Chesapeake Bay on the west, the State of Delaware on the east, Sassafras River on the north and Chester River on the south, is practically surrounded by tidal water except on its eastern boundary. Tides within the bay, along the shores of this county, have an extreme range of 1.8 feet at the mouth of Chester River; 1.4 feet at Tolchester Beach; and 1.6 feet at Howell Point. In Chester River the head of tide is near Millington and the extreme range at Holton Point is 2.0 feet; at Melton Point is 2.1 feet; at Chestertown is 2.3 feet. In Sassafras River the head of tide is near Sassafras and the extreme range is 2.3 feet at Betterton and 2.7 feet at Frederick.

The streams within the county are tributary to Chesapeake Bay either directly or through Sassafras and Chester rivers. The largest of these tributary streams is Cypress Branch which drains an area of 38 square miles, part of which lies in Delaware. The next largest is Morgan Creek which drains an area of 33 square miles. The minimum flow of these streams is of course small, but the slopes may be considerable, appearing to be as great as twenty feet per mile in some instances. The total fall on any stream cannot be great, however, as no land in the county reaches an elevation of more than 100 feet above sea level.

No measurements of discharge of these streams have been made.

The Census Office has listed nine small water powers which have been utilized for grist and flour mills, as follows:

Name	Postoffice	Wheels	Horse-power
Trinks, Henry	Galena	1	25
Spear, E. W.	Millington	1	50
Higman, J. E.	Millington	1	35
Dreka, L. H.	Sassafras	1	20
Price and Topping	Chestertown	2	22
Woodall, A.	Galena	1	35
McKnett, H. W.	Kennedyville	2	35
Cooper, Mrs. H. E.	Norton	1	10
Plummer, B. C.	Stillpond	1	8

Rainfall records show an average annual precipitation of about 43 inches.

Place	Length of record	Average annual precipitation
Chestertown	19 years	42.85
Coleman	28 "	42.35
Rock Hall	12 "	41.17
Millington	27	43.39

THE MAGNETIC DECLINATION OF KENT COUNTY

BY

L. A. BAUER

INTRODUCTORY.

The values of the magnetic declination of the needle, or of the "variation of the compass" as observed have been made by the Maryland Geological Survey, the United States Coast and Geodetic Survey, and the Carnegie Institution of Washington at various points within the county are given in Table I.

TABLE I. MAGNETIC DECLINATIONS IN KENT COUNTY.

Stations	Latitude N.	Longitude W. of Greenwich	Date when observed	Value observed	Magnetic Declinations (West)				Observer
					Reduced to				
					1900.0	1905.0	1910.0	1915.0	
	° '	° '		° '	° '	° '	° '	° '	
Chestertown, Court House	39 13.0	76 05.0	1897.4	5 50.0	6 00	6 18	6 42	7 05	L. A. Bauer, Md. G. S.
Chestertown, College 1897	39 13.0	76 04.4	1897.4	5 47.0	5 57	6 15	6 39	7 02	L. A. Bauer, Md. G. S.
Chestertown, College 1908	39 13.0	76 04.4	1908.8	6 32.9	5 58	6 16	6 40	7 03	C. C. Stewart, C. I. of W.
Tolchester	39 12.9	76 14.3	1897.4	5 37.1	5 47	6 05	6 29	6 48	L. A. Bauer, Md. G. S.
Massey	39 18.5	75 48.5	1896.7	6 25.0	6 37	6 55	7 19	7 40	L. A. Bauer, Md. G. S.
Betterton	39 21.9	76 03.9	1899.5	4 03.9	4 06	4 24	4 48	5 09	L. A. Bauer, Md. G. S. and C. & G.S.

Explanations: The date of observation is given in years and tenths of; January 1, 1900, would accordingly be expressed by 1900.0 and similarly with regard to January 1, 1905 or 1910. See Table II.

For a general description of the methods and instruments used, reference must be made to the "First Report upon Magnetic Work in Maryland" (Md. Geol. Survey, vol. i, pt. v, 1897). In the Second

Report (Md. Geol. Survey, vol. v, pt. i, 1905), the various values collected were reduced to January 1, 1900. They are given now also for January 1, 1905 and 1910. Some slight changes have been made in the previously published values. The First Report contains an historical account of the phenomena of the compass needle and discusses fully the difficulties encountered by the surveyor on account of the many fluctuations to which the compass needle is subject. To these reports the reader is referred for any additional details.

MERIDIAN LINE

In compliance with the instructions from the County Commissioners, dated April 15, 1897, a true surveyor's line was established by L. A. Bauer of the Maryland Geological Survey on May 29, 1897, on the Court House grounds at the County seat of Chestertown. Owing to the lay and character of the grounds, the monuments had to be set on a true northeast-southwest line, instead of a true north and south line. Approved astronomical methods were used and the line may be taken to be correct within one minute. An official report containing all necessary information was furnished for the Court House files.

The monuments marking the line are granite posts 6x6 inches square and 4 feet long; they are imbedded in several courses of concrete and were allowed to project about 5 inches above the ground. They were planted so that the letters on the monuments (N M on the S W stone and S M on the N E stone) would indicate approximately the true north and south. In each monument there was leaded and countersunk a one-inch brass bolt, 3 inches long; the line passing through the center of the crosses cut in the brass bolts is the *true northeast-southwest line*. The year 1897 appears on each stone, the northeast one being 103.7 feet from the northeast corner of the Court House and 121.4 feet from the northwest corner.

DESCRIPTION OF STATIONS

Chestertown, Court House, 1897.—At the northeast stone of the true surveyor's line established on the Court House grounds in 1897.

Chestertown, Washington College, 1897.—On the campus of the College grounds. This being only an auxiliary station, it was not permanently marked.

Chestertown, 1908.—Same as L. A. Bauer's station of 1897. In the southeastern part of grounds of Washington College, 130 feet (39.6 meters) north of the south edge of the grounds, 97 feet (29.6 meters) northeast of an elm tree, and 123.2 feet (37.6 meters) from the corner of a board fence enclosing the field of Mr. White. Station is marked by a blue marble post 5 by 5 by 24 inches (12.7 by 12.7 by 61 cm.) lettered on top "C.I.1908" and sunk flush with the surface of the ground. The period after the letter "I" marks the exact point. The following true bearings were determined: cross on Catholic Church, 27° 26'.8 West of South; cross on Methodist Protestant Church, 8° 15'.5 West of South; cross on Methodist Episcopal Church, 1° 45'.6 West of South.

Tolchester, 1897.—In the race-track back of the picnic grounds.

Massey, 1896.—On the north side of road to Clayton near small school house, about one mile from railroad station; 121 feet north-northwest of corner of school house.

Betterton, 1899.—On the hill west of Bettertown Hotel owned by Mr. John Henry Crew. Precise spot is in line with chestnut tree, on the northeast side of the hill, and the northeast corner of Mr. Crew's house, about one-third of the way from said tree. Point marked by a wooden peg.

With the aid of the figures in Table II the surveyor can readily ascertain the amount of change of the needle between any two dates. For practical purposes it will suffice to regard the change thus described as the same over the county. It should be emphasized, however, that when applying the quantities thus found in the re-running of old lines, the surveyor should not forget that the table cannot attempt to give the correction to be allowed on account of the error of the compass used in the original survey.

TABLE II.

SHOWING CHANGE IN THE MAGNETIC DECLINATION AT CHESTERTOWN FROM 1700 TO 1915.

The following table is reproduced from page 482 of the First Report cited above except for the extension to 1915.

Year Jan. 1	Needle pointed	Year Jan. 1	Needle pointed	Year Jan. 1	Needle pointed	Year Jan. 1	Needle pointed
	° '		° '		° '		° '
1700	6 10 W	1750	3 16 W	1800	1 09 W	1850	2 50 W
05	5 59 W	55	2 56 W	05	1 08 W	55	3 09 W
10	5 47 W	60	2 38 W	10	1 10 W	60	3 29 W
15	5 33 W	65	2 22 W	15	1 14 W	65	3 48 W
20	5 15 W	70	2 04 W	20	1 21 W	70	4 09 W
25	4 58 W	75	1 48 W	25	1 31 W	75	4 28 W
30	4 38 W	80	1 35 W	30	1 43 W	80	4 49 W
35	4 18 W	85	1 24 W	35	1 58 W	85	5 08 W
40	3 58 W	90	1 16 W	40	2 16 W	90	5 26 W
45	3 38 W	95	1 12 W	45	2 32 W	95	5 44 W
1750	3 16 W	1800	1 09 W	1850	2 50 W	1900	5 58 W
						05	6 16 W
						10	6 40 W
						1915	7 03 W

The declination is westerly over the entire county and is increasing at an average annual rate of about 5 minutes.

To reduce an observation of the magnetic declination to the mean value for the day of 24 hours, apply the quantities given in the table below with the sign as affixed:

Month	6 A.M.	7	8	9	10	11	Noon	1	2	3	4	5	6 P.M.
January	−0.1	+0.2	+1.0	+2.1	+2.4	+1.2	−1.1	−2.5	−2.6	−2.1	−1.3	−0.2	+0.2
February	+0.6	+0.7	+1.5	+1.9	+1.4	−0.1	−1.5	−2.1	−2.5	−2.0	−1.2	−0.8	−0.4
March	+1.2	+2.0	+3.0	+2.8	+1.6	−0.6	−2.5	−3.4	−3.7	−3.3	−2.3	−1.2	−0.5
April	+2.5	+3.1	+3.4	+2.6	+0.8	−2.1	−4.0	−4.1	−4.2	−3.6	−2.3	−1.2	−0.2
May	+3.0	+3.8	+3.9	+2.6	+0.1	−2.4	−4.0	−5.0	−4.5	−3.6	−2.3	−0.9	+0.1
June	+2.9	+4.4	+4.4	+3.3	+1.1	−2.0	−3.6	−4.5	−4.5	−3.8	−2.6	−1.2	−0.2
July	+3.1	+4.6	+4.9	+3.9	+1.8	−1.2	−3.4	−4.4	−4.7	−4.2	−2.8	−1.3	−0.3
August	+2.9	+4.9	+5.4	+3.7	+0.4	−2.8	−4.7	−5.1	−4.9	−3.7	−1.9	−0.6	+0.3
September	+1.8	+2.8	+3.4	+2.5	+0.3	−2.7	−4.4	−4.6	−4.2	−4.0	−1.4	−0.3	−0.1
October	+0.5	+1.6	+3.1	+2.8	+1.4	−1.0	−2.7	−3.3	−3.4	−2.4	−1.3	−0.4	−0.4
November	+0.5	+1.2	+1.7	+1.8	+1.1	−0.5	−2.0	−2.7	−2.6	−1.8	−1.0	−0.2	+0.2
December	+0.2	+0.3	+0.8	+1.8	+1.8	0.0	−1.6	−2.4	−2.3	−1.8	−1.1	−0.3	+0.1

ANGLE.

At the northeast stone the angle between the true northeast-southwest line and the northeast corner of the Court House is 17° 34' and for the northwest corner of the Court House the angle is 33° 41'.

The latitude of the Court House may be taken to be 39° 13.0', and the longitude 76° 04.4' W of Greenwich or 56' East of Washington. To obtain true local mean time, or solar time, subtract from Eastern or Standard time 4 minutes and 26 seconds.

FIG. 1.—VIEW SHOWING LOBLOLLY PINE SAPLINGS.

FIG. 2.—LOBLOLLY PINE FOREST NEAR ROCK HALL. AT ITS NORTHERN LIMIT OF GROWTH IN THE UNITED STATES.

THE FORESTS OF KENT COUNTY

BY

F. W. BESLEY

INTRODUCTORY.

Kent is pre-eminently an agricultural county that has reached a high state of farm development. There is a small percentage of woodland, rather unequally distributed, but inasmuch as the forested areas have been brought to a nearly irreducible minimum, forest products have a higher value for local uses than obtains in other counties where there is a larger percentage of woodland. This report is based upon a complete forest survey, made in 1907, by the State Department of Forestry, an established agency for the investigation of forest conditions, and prepared to give reliable information and advice concerning forest management to all woodland owners. The data upon which the report is based has been revised, since the original survey was made, in order to bring it more nearly up to date.

The land area of the county consists of 179,872 acres classified as follows:

```
Improved farm land........  132,726 acres, or 73%
Woodlands ................   33,776 acres, or 19%
Marsh lands..............     7,000 acres, or  4%
Waste land ..............     6,370 acres, or  4%
```

There is a considerable variation in the agricultural lands, in different parts of the county, giving rise to a variety of crops, but most of it is of a good quality of clay loam that produces excellent yields of corn and wheat. The demand for farm lands has resulted in a reduction of the forest areas to a point where practically all tillable land has been cleared, so that the present wooded area will, probably, always remain. Indeed, during the past few years of agricultural depression a considerable acreage of land, formerly cultivated, has reverted and is now classed as waste land.

At the time of settlement magnificent mature hardwood forests covered almost the entire county. The only non-forested areas were the salt marshes and a few grassy glades where tree growth could not successfully compete with the grasses. No sooner had settlement begun and the necessity for cleared land on which to grow food crops asserted itself, than the natural conditions were changed and the forests destroyed. With the increase of population the clearing of land went forward with greater rapidity. Timber was overabundant and consequently of very little value. The land was found very productive and nearly all of it so situated as to be easily tillable. This course, consistently followed for over 200 years could lead to but one result, the forests fell before the ax of the settler and of the farmers who flocked to the land of great agricultural promise, and today there exists a high state of agricultural development, but as for lumber and other construction material, the county is a heavy importer having long ago ceased to produce enough timber for the local demand. The scarcity of timber is not entirely due to the clearing away of the forests for farm crops, though that is the chief cause. There is still left nearly 19 per cent of the total land area in forest which, if properly managed and fully productive, would supply at least three-fourths of the local needs. The present difficulty and the problem for serious consideration is that the present wooded area has been so mismanaged that it is not producing one-third of a full crop. The forests for many years have been subjected to a system of culling in which the best trees of the best species are being constantly taken with the result that the forests have not only been thinned to the point where they are only partially stocked, but the trees that constitute the present stands consist largely of scrubby defective specimens and those of inferior species which, because of their worthlessness, were left in

the woods and are now simply encumbering the ground and prevent-
ing a more valuable growth.   A radical change in the method of
handling the forests is imperative to put all of the lands, the woodlands as well as the fields, in a state of highest productiveness and
demands the combined efforts of the farmer and the forester.

The present forest resources of the county are graphically shown
by the following tabular statement:

TABLE I.

WOODED AREA, STAND, AND VALUE OF TIMBER BY ELECTION DISTRICTS.

District	Total area of district —acres	Total wooded area—acres	Percentage of woodland	Total stand of saw timber—M ft.	Approximate stumpage value of saw timber
Massey I	43,405	10,409	24	19,379	$167,080
Kennedyville II	42,374	4,842	11	10,558	95,008
Worton III	26,419	4,480	17	7,890	61,264
Chestertown IV	5,376	608	11	1,382	12,560
Rockhall V	19,059	4,029	21	6,201	50,112
Fairlee VI	23,642	4,835	20	10,128	90,224
Pomona VII	19,597	4,615	24	8,097	68,536
Totals	179,872	33,818	19	63,135	$544,784

The value of saw timber given in the table represents its value
as it stands in the tree in the woods without any labor expenditure.
The same timber after cutting and sawing would represent a value
at the mills of about $1,750,000.   The table shows that the wood-
lands of the county are not evenly distributed, two districts having
each 24 per cent of wooded area and two others but 11 per cent each.
Massey district, which has the largest land area, has more than
twice as much woodlands as any other district and also has the
largest amount of standing timber.   It is here that some of the
largest lumbering operations in the county are in progress.   Kennedyville and Chestertown districts have the smallest per cent of
woodlands but in point of stand and value of timber Kennedyville,
which is the second largest district in the county, ranks among the
first.   The other districts, viz., Worton, Rock Hall, Farlee, and

Pomona each have nearly the same amount of woodlands, though the percentage of wooded area differs in each, due to variation in the relative size of the districts.

THE CHARACTER OF THE WOODLANDS

Kent County is the northernmost county of the Eastern Shore Peninsula entirely within the tidewater section and in consequence some of the tidewater tree species attain here the northern limit of their distribution. The most notable example is the loblolly pine, a valuable timber tree of the south, forming extensive forests in all of the lower counties where it is the principal timber tree. Loblolly pine forests, however, are not found farther north than the southern part of Kent County. There is likewise in this transition zone a curious mingling of the northern and southern species of trees. The forest vegetation is most luxuriant and, for the botanist, presents an interesting field for study.

The woodlands are largely confined to the poorly drained soils along the water courses or to the short abrupt slopes adjacent to the Sassafras River and its tributaries and along the bay shore. In undrained soils there are relatively few kinds of trees that will thrive, such as red gum, black gum, red maple, pin oak, willow oak, etc., and these are generally of less value than upland species. On exposed slopes along the bay shore and the Sassafras River the conditions are not favorable for the best tree growth, but in such locations they serve their most useful purpose in protecting the short abrupt slopes from soil erosion, as well as affording an excellent windbreak against the cold northwest winds. The principal species found in such locations are chestnut oak, spanish oak, chestnut, and locust.

The area, stand, and value on the stump of the saw timber of the different classes in the several districts is shown by the following table:

FIG. 1.—VIEW OF DESTRUCTION OF A THRIFTY STAND OF YOUNG TIMBER BY FIRE—THE WORST ENEMY OF THE FOREST.

FIG. 2.—VIEW SHOWING POLES FOR FISH POUNDS AT ROCK HALL. STRAIGHT SPRUCE PINE TREES ARE THE ONES GENERALLY USED.

TABLE II.

DISTRIBUTION OF WOODLANDS, STAND AND VALUE OF SAW TIMBER
BY ELECTION DISTRICTS.

District	Merchantable Hardwoods			Merchantable Pine		
	Area acres	Stand M Bd. Ft.	Value	Area acres	Stand M Bd. Ft.	Value
Massey I	1,130	4,520	$45,200	376	1,504	$15,040
Kennedyville II	1,318	5,272	52,720	.....	.....	.....
Worton III	268	1,072	10,720	.....	.....	.....
Chestertown IV	188	752	7,520	.....	.....	.....
Rock Hall V	.....	.....	.....	63	252	2,520
Fairlee VI	1,150	4,600	46,000	.....	.....	.....
Pomona VII	470	1,880	18,800	.....	.....	.....
Totals	4,524	18,096	$180,960	439	1,756	$17,560

District	Culled Hardwoods			Mixed Hardwoods and Pine		
	Area acres	Stand M Bd. Ft.	Value	Area acres	Stand M Bd. Ft.	Value
Massey I	8,830	13,245	$105,960	73	110	$880
Kennedyville II	3,480	5,220	41,760	44	66	528
Worton III	4,212	6,318	50,544	.....	.....	.....
Chestertown IV	420	630	5,040	.....	.....	.....
Rock Hall V	3,552	5,328	42,624	414	621	4,968
Fairlee VI	3,685	5,528	44,224	.....	.....	.....
Pomona VII	4,145	6,217	49,736	.....	.....	.....
Totals	28,324	42,486	$339,888	531	797	$6,376

THE FOREST TYPES

There are three types of forest in the county which may be
designated as mixed hardwood, mixed hardwood and pine, and pure
pine. The mixed hardwood type is the most important since it
covers 97 per cent of the wooded area and comprises nearly the
entire stand of timber. The mixed hardwood and pine type occurs
in three districts—Massey, Kennedyville, and Rock Hall—the latter
district containing the bulk of it. This type covers about 531 acres
and represents less than 2 per cent of the total wooded area of the
county. The pure pine type occurs in but two districts, Massey and
Rock Hall. That in the Massey district is spruce pine, while that
found in Rock Hall district is part spruce pine and part loblolly

pine. The latter is the more valuable of the two as a timber tree but of small acreage as compared with spruce pine. The entire area of the pure pine type covers but 439 acres and represents but a little over 1 per cent of the total woodlands of the county.

MIXED HARDWOOD TYPE.

The mixed hardwood type, comprising the greater bulk of the wooded area, consists of a variety of species, often as many as twenty-five different kinds of trees occurring on a single acre. The more valuable species, such as white oak, hickory, black oak, and red gum were originally much more abundant. In late years these species have been cut rather closely and there is, in consequence, an increasing proportion of the less valuable species, such as black gum, red maple, beech, pin oak, and the dense underbrush of dogwood and iron wood to usurp the ground to the exclusion of a more valuable second growth. Since the wooded areas are largely confined to the short, abrupt slopes toward the creeks on the bay shore and to the undrained situations growth is, on the whole, slow. On the better drained soils, however, growth is rapid and the species represented are largely of the valuable kinds that make good timber.

For the purpose of greater accuracy in estimating the stand of saw timber and for the sake of a better representation of actual conditions the mixed hardwood forests were divided into three classes based upon the average stand of saw timber per acre. Two classes are designated, merchantable hardwoods and culled hardwoods. (See Table II.)

The merchantable hardwood class includes the stands containing sufficient saw timber to warrant logging operations, and represents 20 per cent of the wooded area. The average stand of saw timber of this class is 4,000 feet, board measure, per acre, on the 4,524 acres in the county, and gives a total stand of 18,096,000 feet, valued at $180,960 on the stump.

The culled hardwood class represents stands that have either been severely culled or repeatedly cut over until the amount of saw timber will average only 1,500 feet to the acre. This type covers 28,324 acres or nearly 76 per cent of the total wooded area. The stand of saw timber is 42,486,000 feet, with a stumpage value of $339,888.

MIXED HARDWOOD AND PINE TYPE.

This type of forest consists of a mixture of hardwood and pine and is represented in Table II. It occurs in three districts of the county, namely, Massey, Kennedyville, and Rock Hall. In Massey and Kennedyville districts, in the eastern part of the county, it covers 117 acres and the pine, in mixture with the hardwoods, is exclusively the spruce pine. In Rock Hall district, in the south-western part of the county where this type covers 414 acres, the mixture consists of spruce pine with hardwoods in some cases, and of loblolly pine with hardwoods in other cases. Most of these mixed stands were originally a pure hardwood growth where, in the process of culling, sufficient open places were created for the pine to become established by seeding from neighboring pine stands. The mixed hardwood and pine forests are found on 1½ per cent of the woodlands of the county and have a total stand of 797,000 feet, board measure, with a stumpage value of $6,376.

PURE PINE TYPE.

This type is found in but two districts in the county, namely, Massey in the eastern part, and Rock Hall in the southwestern part. In Massey district the pine is spruce pine entirely and covers 376 acres. Spruce pine does not make large timber and, because it is usually knotty and small, it seldom makes a better grade than box board lumber or scantling. In Rock Hall district there are 63 acres of pure pine stands, the most of which are loblolly found on the poorer drained situations. Lower Kent County is the northern

limit of range of loblolly pine forests on the Eastern Shore Peninsula. The total stand of pine in the 439 acres of this class is
1,756,000 feet, board measure, and represents a stumpage value of
$17,560.

THE NATIVE TREES

Fifty-eight native species of tree size, together with five introduced species that have become common, are found in the county.
The following is a complete list. The first five are coniferous (ever-
greens), the remainder are hardwoods (deciduous):

CONIFERS.

Botanical Name	Common Name
Juniperus virginiana L.	Red Cedar
Pinus echinata Mill.	Short Leaf Pine
Pinus rigida Mill.	Pitch Pine
Pinus taeda L.	Loblolly Pine
Pinus virginiana Mill.	Spruce Pine

HARDWOODS.

Acer rubrum L.	Red Maple
Alnus maritima Nutt.	Swamp Alder
Amelanchier canadensis Med.	Service Berry
Aralia spinosa L.	Hercules Club
Asimina triloba Dunal.	Paw Paw
Betula nigra L.	River Birch
Carpinus caroliniana Walt.	Blue Beech
Castanea dentata Borkh.	Chestnut
Celtis occidentalis L.	Hackberry
Cercis canadensis L.	Redbud
Cornus florida L.	Flowering Dogwood
Crataegus coccineae	Scarlet Thorn
Diospyros virginiana L.	Persimmon
Fagus grandifolia Ehrh.	Beech
Fraxinus americana L.	White Ash
Fraxinus pennsylvanica Marsh.	Red Ash
Hamamelis virginiana L.	Witch Hazel
Hicoria alba L.	White Hickory
Hicoria glabra Mill.	Pignut Hickory
Hicoria minima Marsh.	Bitternut Hickory
Ilex opaca Ait.	Holly
Juglans cinerea L.	White Walnut
Juglans nigra L.	Black Walnut

Liquidambar styraciflua L..............Red Gum
Liriodendron tulipifera ....Yellow Poplar
Magnolia virginiana L................Sweet Bay
Morus rubra L........................Red Mulberry
Myrica cerifera L.....................Wax Myrtle
Nyssa sylvatica ˙Marsh................Black Gum
Platanus occidentalis L................Sycamore
Prunus pennsylvanica L................Fire Cherry
Prunus serotina Ehrh.................Wild Black Cherry
Prunus virginiana L...................Choke Cherry
Quercus alba L.......................White Oak
Quercus coccinea Muench..............Scarlet Oak
Quercus lyrata Walt..................Overcup Oak
Quercus marilandica Muench..........Black Jack Oak
Quercus michauxii Nutt ....Basket Oak
Quercus stellata Wang................Post Oak
Quercus nigra L......................Water Oak
Quercus palustris Muench.............Pin Oak
Quercus phellos L.....................Willow Oak
Quercus bicolor Willd.................Swamp White Oak
Quercus borealis maxima Ashe..........Northern Red Oak
Quercus rubra L......................Southern Red Oak
Quercus velutina Lam.................Black Oak
Rhus typhina L.......................Staghorn Sumach
Robinia pseudacacia L.................Black Locust
Salix discolor Muhl...................Glaucous Willow
Salix nigra Marsh.....................Black Willow
Sassafras sassafras Karst.............Sassafras
Ulmus americana L....................White Elm
Ulmus fulva Michx....................Slippery Elm

INTRODUCED SPECIES THAT HAVE BECOME COMMON.

Acer negundo L.......................Ash-leaved Maple
Ailanthus altissima Swing.............Ailanthus
Gleditsia triacanthos L................Honey Locust
Populus alba L.......................Silver Poplar
Toxylon pomiferum Rafn..............Osage Orange

IMPORTANT TREE SPECIES

Saw timber is so scarce in the county that practically all species are cut and used for building material. There are, however, several species that by their abundance and good qualities are regarded as of special importance.

WHITE OAK.

This produces the highest priced construction timber of any tree in the county, and is becoming scarce. It is in great demand for bridge plank, cross-ties, piling, and general construction purposes. The tree is found in moist, deep soils associated with other oaks and beech. The demand for white oak timber is so great as to force the cutting of trees before they are mature and results in lessening the proportion of this species in the forest stands.

SPANISH OAK.

Is one of the most common trees on well-drained soils. It does not rank in value with the white oak but is its most important substitute and is largely used for building material. It is used to some extent for railroad ties and piling.

WILLOW OAK AND PIN OAK.

Are abundant in the poorly drained soils along the streams. The wood is not so valuable as that of the other oaks but it is largely used for rough lumber for construction purposes where exposure to weather is not required. Since the trees are usually straight with slight taper they make good piles.

RED GUM.

This species of late years has come into general use for cutting into veneer, which is manufactured into berry boxes and fruit and vegetable baskets. It is abundant and makes a good growth on low ground, where it competes with black gum and red maple, though it frequently occurs in almost pure, even-aged stands.

YELLOW POPLAR.

Is one of the most valuable woods in the county. The tree attains large size in the deep forest soils along streams, but repeated culling has so reduced its chance for survival as to make it a comparatively rare tree.

PINE.

Is an important tree in the southern and southeastern parts of county. Of the two species of pine in the county, spruce pine and loblolly, the former is much more abundant though the timber is of less value. The principal use of spruce pine is for fire wood and pulpwood while loblolly pine is in demand for lumber, especially for box boards for which it is particularly adapted. Both pines are slightly increasing in area, the spruce pine occupies the better drained light soils, while the loblolly is generally found on low sandy grounds, adjacent to swamps.

LUMBER AND TIMBER PRODUCTION

LUMBER.

The available saw timber is in such small tracts and so much scattered that only small operations are possible. There were nine saw mills operating in the county in 1925, most of them running but a few months and cutting for local orders. Practically all species are cut for lumber but the oaks and pines are the most important. A considerable amount of dead chestnut was utilized, but most of that left has been dead so long that it is unfit for saw timber.

RAILROAD TIES.

There are about 30 miles of railroad in the county which draw heavily upon the local timber supply for railroad ties. Oak is used almost exclusively, the white oak being in greater demand. The normal annual requirements are about 15,000 ties for replacement and new construction. The supply of available tie timber has been so far depleted that imported treated ties are now being generally used.

POLES.

The network of telephone and telegraph lines, extending through the county, require a large number of poles for maintaining existing lines and constructing new ones. Chestnut continues to be the species almost universally used for the purpose, and before the chestnut blight appeared, there was an abundance of pole timber of this species. In recent years, however, with the practical destruction of all standing chestnut timber, the pole line companies are beginning to import heavily to supply local needs.

FENCING MATERIAL.

A large quantity of fence posts are required annually, as would naturally be expected in a county where dairying and stock raising are so important. Practically all fields are fenced, since under the existing practice of crop rotation, nearly every field is used for pasture periodically. It is estimated that 85,000 fence posts are required annually for replacement and for new fence lines. Red cedar and black locust make the best fence posts and are generally used. As a third choice, a great deal of chestnut was formerly used, but the supply has been, practically, destroyed by the chestnut blight.

Indications are that within a few years, the use of treated fence posts will be required to supplement the diminishing supply of naturally durable woods. It has been demonstrated at the Maryland Experiment Station that cheap woods of low natural durability, such as pine or gum, can be made to last for 20 years or more, when properly treated with creosote at a reasonable expense.

FUELWOOD.

The quantity of wood used annually for fuel is greater than that for all other uses combined. Notwithstanding the increasing amount of coal used, the large majority of people depend upon wood for fuel. Most farmers have woodlots from which they secure their fuelwood, as well as the material for fencing and construction purposes around the farm. The farmer who has as much as 15 acres of woodland can secure the needed fuel by utilizing the dead and defective trees in the nature of thinnings and improvement cuttings,

FIG. 1.—VIEW SHOWING MISMANAGED STAND OF HARDWOOD NEAR HOWELL'S POINT.
WORTHLESS TREES SHOULD BE ELIMINATED BY CUTTING.

FIG. 2.—VIEW SHOWING FUEL WOOD CUT FROM THINNINGS IN A LOBLOLLY PINE THICKET.

thus maintaining the productivity of his forest lands, while utilizing them for immediate needs.

The wood and timber taken from the forest each year is much greater than the annual growth. The result is a constant depletion which has already made the county a heavy importer of building material.

This emphasizes strongly the need of conserving and increasing timber production to meet home needs which cannot long be supplied from the surplus of neighboring counties, where supplies are being rapidly depleted. The universal adoption of systematic forest management in the county would increase the production of the forests certainly to three times their present output and probably more.

WOOD USING INDUSTRIES

In addition to the nine saw mills, there are wood-using industries operating in the county, which convert rough lumber, or veneer logs, into manufactured products. There are not less than three such establishments, employing 85 men, which, annually, convert about 2,000,000 board feet of rough sawed lumber into manufactured products, principally boxes, baskets, crates, flooring, ceiling, window and door frames and other interior finish.

These are growing industries which represent considerable capital and give employment to a great many people. It is important that these industries be encouraged and extended and that can only be done by maintaining the supply of timber, which is the raw material, upon which they must depend. The woodland owner, who is the timber producer, is likewise helped in securing a good local market for his timber. The county is also enriched by retaining this source of wealth for the benefit of its own people.

The principal species used are yellow pine, red gum, yellow poplar, American elm, cypress, and a number of species of oak. Of the amount used approximately half of it is Maryland grown, the remainder coming from outside the State.

FOREST MANAGEMENT

In view of the fact that the local timber supply is not equal to the demand and that prices are certain to advance considerably, the question is, or should be, how can these woodlands be made more productive? A study of the conditions has shown that greatly increased timber production is possible and can easily be accomplished by applying certain well known principles of forest management.

1. Full timber production is only possible where the woodland is fully stocked with growing trees, that is, where no open places occur in the forest.

2. The highest yield, quantity and quality considered, can only be obtained by encouraging the species best adapted to the location, and those for which there is the greatest demand, and at the same time weeding out the undesirable kinds.

3. The forest must be fully protected against fires, grazing, and tree diseases.

In a mixed hardwood stand it is not often possible to attain the ideal conditions, but the nearer such conditions are approached the better will be the results.

In the first place the number of trees required to make a fully stocked stand will depend upon the age of the stand, the fertility of the soil and the species, hence no definite number can be given. A stand is, however, fully stocked no matter what may be the age, species, or locality, when the tops of the trees are so close together that the branches of each tree touch, or nearly touch, on all sides the branches of its neighboring trees. If the trees are so close that the branches interlace, they are too close to make the best growth and a thinning is needed. In the young stages of growth a little crowding is beneficial, as it forces the trees to shoot up rapidly for light, and at the same time the lower branches are killed by shading and drop off. When the main height growth is attained less shading on the sides is required for then diameter increase is most needed, and can be encouraged by giving the tree more room.

On good soil there will be fewer trees of a given age per acre than on a poor soil, although on the former the trees will be larger. Then, too, some species will grow in dense stands while others will not. White oak, beech, and spruce pine will grow so close together as to completely shade the ground, while locust, poplar, and black oak require more room in which to develop.

In the second place such species as white oak, red oak, yellow poplar, hickory, and red gum are of the greatest commercial value and should be encouraged at the expense of the less valuable species. This can be done by improvement cuttings made at proper times and in connection with thinnings when required. The woods should be gone over frequently to remove dead and suppressed trees and those of inferior kinds that are crowding the more valuable species named above. Such material can be utilized by the farmer for fuel and the more valuable timber growth reserved to supply the future market demands at a good price.

When the time comes to cut a stand of timber and the land is to be held for further wood production, the main question should be how to secure a valuable new growth. The method of cutting will be a controlling factor. Nearly all hardwood sprout from the stump and if the trees are cut while in full vigor the production of sprouts will be abundant. Under the usual practice the valuable trees have been removed and the less valuable ones, for which there is little market demand, have been left. This repeated culling has brought about a radical change in the representation of species and greatly reduced the producing capacity of the woodlands. In order to restore normal conditions it will be necessary to adopt different methods. The particular method to be adopted will depend upon conditions present and cannot be stated in definite terms. In the care of the woodlot where there is a constant need for fuel, the inferior species can gradually be thinned out as more room is needed for the better trees and in that way much of the present inferior growth will give way to the more valuable trees.

MANAGEMENT OF MIXED HARDWOOD AND PINE.

Forest management for this type of forest will be similar to that recommended in the case of the mixed hardwood stands, since the hardwoods constitute at least 75 per cent of the stand. If a larger percentage of reproduction of pines is desired in the growth, it will be necessary to leave a number of pine seed trees per acre for re-stocking as pines do not sprout from the stumps, as do the hardwoods. The pine will only have a chance to succeed where there are open places, since the hardwood sprouts will generally grow faster than pine seedlings and kill out most of those that succeed in getting started. The fact that the pine is increasing in distribution is due to the repeated culling of the hardwood forests, thereby creating open spaces where the light pine seed finds a chance to germinate and grow. Were it not that the pine is a prolific seeder and that the seed is blown great distances by the wind it would in time be almost exterminated by the more persistent hardwood growth. In dealing with the spruce pine which, because of its slower growth and less valuable product, is not as desirable as the hardwoods, the plan of management would be to reduce the representation of that species in the mixture. This can be best accomplished by thinnings and improvement cuttings. When the mixed stand is in need of thinning the spruce pine should be regarded as of lesser value or even a tree weed, and sacrificed whenever it interferes with the development of a more valuable hardwood, hence the operation is called an improvement cutting. Since pine does not sprout from the stump one cutting will suppress the tree. On the other hand, in dealing with loblolly pine which is a rapid growing tree reaching large saw timber size and therefore to be encouraged in the mixed forest, the operation would be reversed, that is, instead of making improvement cuttings to eliminate the tree, its competitors should be cut away as fast as they threaten to suppress it or interfere with its proper development.

MANAGEMENT OF PINE STANDS.

Pine nearly always grows in even-aged stands and therefore the clear cutting system is the one to follow. The time to cut is when the trees have reached financial maturity and that is in reality the time when they will bring the best new returns—taxes, interest, and rate of growth considered. Spruce pine, which is the pine most largely represented in the county, grows very slowly after it gets to be six or eight inches in diameter and ordinarily should be cut about that time, as it rarely reaches good saw timber size and when it does remain long enough for saw logs, the taxes and interest, or rental, of high priced land more than offset its stumpage value. Its chief market is for cordwood, pulpwood, or mine props, which does not require large sizes.

Usually the best system of handling pure stands of spruce pine is to cut them clean at maturity and plant with loblolly pine or some other species, since spruce pine is not a tree that is profitable to grow.

The loblolly pine stands in the southwestern part of the county, while smaller in extent are important because loblolly pine is one of the most valuable trees in the county. The lumber is not so valuable as the white oak, but it grows so much faster that the money yield in a given time is considerably greater. It does not reach financial maturity until it gets to be about 15 inches in diameter on the stump when it will be from 35 to 50 years old. At that size it makes good saw timber or mine props.

Where loblolly pine occurs in pine stands and has reached maturity, the best method of cutting is to cut clear with the exception of three or four good seed trees to the acre well distributed in order to insure a re-stocking of this species. Where there is a heavy hardwood undergrowth mixed with the pine, the competing hardwoods will have to be cut back in order that the pine may succeed as the more important species.

TREE PLANTING

There are small open areas throughout the county, not now utilized for field crops that can be profitably employed in growing timber. There are a number of tree species of rapid growth and high value, which can be profitably employed for the purpose. Black locust, a rapid growing hardwood making an excellent fence post, is one of the most promising. This species will grow rapidly on good soil, and does well even on poor soil. The increasing demand for fence posts, and its adaptability for planting on waste lands on the farm, make it especially desirable. Planted stands of locust should produce on good soil fence posts in 12 years. Loblolly pine is another species that can be highly recommended for forest plant- ing. It is the most rapid growing of the pines, a native tree in the county, and will produce good saw timber in less time than any other native species. · The cost of starting a plantation will depend upon the cost of the trees and the labor required in planting. This will ordinarily be less than $10.00 per acre.*

FOREST PROTECTION

The most serious enemy of the forest is fire. Where fires are permitted to run through the woods, there can be no satisfactory tree growth. Even a light surface fire causes serious damage in burning up the leaf litter, which is needed to keep the soil in a good physical condition by acting as a mulch to conserve moisture and soil fertility. The seed and small seedlings, intended for re-stocking the forest, are destroyed, young trees are killed or·seriously injured, and even the larger ones are often fire scarred, exposing them to decay. Fortunately, the natural conditions in the county are such

* The State Department of Forestry operates a forest nursery from which stock may be obtained at low cost for forest planting.

as to limit greatly the fire damage. The woodlands are usually in small areas, often located along the water courses, where fires are not apt to occur, but there have been destructive fires, and there is always the fire danger during the dry periods in the spring and fall. Every woodland owner should be on the alert during these periods to prevent fires and ready to fight promptly any that do occur. Where outside assistance is required in combatting fires, the State forest wardens are available, with the authority to employ such assistance as may be required to bring fires promptly under control.

CHESTNUT BLIGHT

The chestnut blight is the most destructive tree disease ever known in this county. It is a fungus disease brought into the United States, probably on nursery stock, from the Orient in the nineties, coming under definite observation in the vicinity of New York in 1904. From this point, it has steadily spread at an average rate of about 40 50 miles yearly. The blight appeared in the county about 1909 and has killed practically every chestnut tree. It is a bark disease that affects only the bark and the layer of wood lying just beneath it. Consequently, if diseased trees are utilized soon after they die and before natural decay starts, the wood is as sound as that from living trees.

Much time and money has been expended in trying to find a remedy for the blight or a method of control, without success. The only sensible thing for the land owner to do is to utilize the chestnut before it becomes worthless.

SUMMARY

1. Nineteen per cent of the area of the county is wooded. The total stand of saw timber is over 63,000,000 feet, board measure, with a stumpage value of about $544,784.

2. Much of the wooded area is in poor condition due to injudicious cutting and lack of proper management, so that the forests

are now producing less than one-third the value of product of which they are capable. The annual cut from the forest is greatly in excess of the annual growth.

3. There are three principal types of forest in the county, namely, mixed hardwood, mixed hardwood and pine, and pure pine. Of these the mixed hardwood type is the most important since it covers 97 per cent of the wooded area and comprises nearly the entire stand of available saw timber.

4. There are no less than 63 species of forest trees in the county, 35 of which are useful timber trees.

5. The principle products of the forests are lumber, shingles, railroad ties, poles, piling, mine props, pulpwood, cordwood, firewood, and fence posts.

6. The wood-using industries of the county convert over 2,000,000 feet of rough lumber into manufactured products including wheelwright stock, vehicles, boxes, crates, flooring, ceiling, window and door frames, and other interior finish. The productiveness of the forest must be maintained to continue these industries which give employment to a great many people.

7. By applying the principles of practical forestry to the management of woodlands the forest yields can be greatly increased and the quality of product much improved.

8. There is not much land not suitable for agricultural crops upon which timber growing would be profitable. Such lands should be planted with trees best adapted to the locality, especially those that promise quick returns such as locust or loblolly pine.

9. The forests by reason of their location, generally on low lands, and their division into small woodlots make them less subject to fires than is the case in other counties; nevertheless they do suffer to some extent from this source.

10. Over-grazing in the woodlot is responsible for a considerable share of the damage that forests now suffer.

INDEX

A

Abbe, Cleveland, Jr., 40, 41.
Agricultural conditions, discussed, 127.
Alexander, J. H., 31, 32.
Alexander, Wm. H., 17.
Aquia formation, 71.
    areal distribution of, 71.
    character of materials of, 72.
    paleontologic character of, 73.
    stratigraphic relations of, 74.
    strike, dip and thickness of, 73.
Areal distribution of Aquia formation, 71.
    of Calvert formation, 74.
    of Magothy formation, 62.
    of Matawan formation, 65.
    of Monmouth formation, 68.
    of Raritan formation, 58.
    of Talbot formation, 83.
    of Wicomico formation, 79.
Artesian waters, 100.

B

Bagg, Rufus M., 40, 42.
Bailey, J. W., 29.
Bassler, R. S., 42.
Bauer, L. A., 18, 157.
Berry, Edward W., 7, 43, 44.
Besley, F. W., 18, 161.
Betterton, precipitation at, 149.
    temperatures at, 140, 142, 144.
Betterton Wharf, section near, 63.
Bibliography, 30.
Bog-iron ore, discussed, 99.
Bonsteel, Jay A., 17, 41, 111.
Boyer, C. S., 42.
Brandywine formation,
    sedimentary record of, 91.

C

Calvert formation, 74.
    areal distribution of, 74.
    character of materials of, 75.
    paleontologic character of, 75.
    stratigraphic relations of, 76.
    strike, dip and thickness of, 76.
Case, E. C., 42.

Character of materials of Aquia formation, 72.
    of Calvert formation, 75.
    of Magothy formation, 62.
    of Matawan formation, 66.
    of Monmouth formation, 69.
    of Raritan formation, 59.
    of Talbot formation, 84.
    of Wicomico formation, 79.
Chesapeake Group, 74.
Chester, Frederick D., 29, 35, 36.
Chestertown, precipitation at, 149, 151.
    snowfall at, 153.
    temperatures at, 140, 142, 144, 146.
Chestnut blight, discussed, 179.
Clark, Wm. Bullock, 26, 27, 28, 37, 38, 39, 40, 41, 42, 43, 44.
Clays, discussed, 97.
Climate, discussed, 131.
Climatological stations in county, 132.
Coleman, precipitation at, 149, 151.
    snowfall at, 153.
    temperatures at, 140, 144, 146.
Columbia Group, 76.
Conifers, 168.
Conrad, J. A., 32.
Conrad, T. A., 34.
Contents, 11.
Cretaceous, discussed, 58.

D

Dall, W. H., 42.
Darton, N. H., 27, 28, 30, 38, 39.
Ducatel, J. T., 26, 27, 28, 31, 32, 33, 34.
Dutton, J. R., 133.
Drainage, 50.

E

Eastman, C. R., 42.
Ehrenberg, C. G., 29.
Elkton clay, 115, 122.
Eocene, discussed, 71.
    sedimentary record of, 90.
Eocene water horizon, 107.

F

Fence timber, 172.
Ferguson, John B., 5.